WHAT IS CALCULUS ABOUT ?

NEW MATHEMATICAL LIBRARY

published by
The Mathematical Association of America

The New Mathematical Library (NML) was begun in 1961 by the School Mathematics Study Group to make available to high school students short expository books on various topics not usually covered in the high school syllabus. In a decade the NML matured into a steadily growing series of some twenty titles of interest not only to the originally intended audience, but to college students and teachers at all levels. Previously published by Random House and L. W. Singer, the NML became a publication series of the Mathematical Association of America (MAA) in 1975. Under the auspices of the MAA the NML will continue to grow and will remain dedicated to its original and expanded purposes.

WHAT IS CALCULUS

ABOUT?

by

W. W. Sawyer

University of Toronto

2

MATHEMATICAL ASSOCIATION
OF AMERICA

Illustrations by Carl Bass

Sixteenth Printing

© Copyright, 1961, by Yale University

Library of Congress Catalog Card Number: 61-6227

Manufactured in the United States of America

Note to the Reader

This book is one of a series written by professional mathematicians in order to make some important mathematical ideas interesting and understandable to a large audience of high school students and laymen. Most of the volumes in the *New Mathematical Library* cover topics not usually included in the high school curriculum; they vary in difficulty, and, even within a single book, some parts require a greater degree of concentration than others. Thus, while the reader needs little technical knowledge to understand most of these books, he will have to make an intellectual effort.

If the reader has so far encountered mathematics only in classroom work, he should keep in mind that a book on mathematics cannot be read quickly. Nor must he expect to understand all parts of the book on first reading. He should feel free to skip complicated parts and return to them later; often an argument will be clarified by a subsequent remark. On the other hand, sections containing thoroughly familiar material may be read very quickly.

The best way to learn mathematics is to *do* mathematics, and each book includes problems, some of which may require considerable thought. The reader is urged to acquire the habit of reading with paper and pencil in hand; in this way mathematics will become increasingly meaningful to him.

The authors and editorial committee are interested in reactions to the books in this series and hope that readers will write to: Anneli Lax, Editor, New Mathematical Library, NEW YORK UNIVERSITY, THE COURANT INSTITUTE OF MATHEMATICAL SCIENCES, 251 Mercer Street, New York, N. Y. 10012.

The Editors

NEW MATHEMATICAL LIBRARY

CONTENTS

CONTENTS

WHAT IS CALCULUS ABOUT?

CHAPTER ONE

What Must You Know to Learn Calculus?

In mathematics, a certain surprising thing happens again and again. Someone poses a simple question, a question so simple that it seems no useful result can come from answering it. And yet it turns out that the answer opens the door to all kinds of interesting developments, and gives great power to the person who understands it.

Calculus is an example of this. Calculus begins with an apparently simple and harmless question, "What is speed and how can we calculate it?" This question arose very naturally round about the year 1600 A.D., when all kinds of moving objects—from planets to pendulums— were being studied. Men were then just starting to study the material world intensively. From that study the modern world has developed, with the knowledge of stars and atoms, of machines and genes, that we have today, for good and for ill. One might have expected the study of speed to have very limited applications—to machinery, to falling objects, to the movements of the heavenly bodies. But it has not been so. Practically every development in science and mathematics, from 1600 to 1900 A.D., was connected with calculus. From this single root, in a most unexpected way, knowledge grew out in all directions. You find calculus applied to the theory of gravitation, heat, light, sound, electricity, magnetism; to the flow of water and the design of airplanes. Calculus enables Maxwell to predict radio twenty years before any physicist can demonstrate radio experimentally; calculus still plays a vital role in Einstein's theory of 1916 and in the new atomic theories

3

of the nineteen-twenties. Apart from these, and many other applications in science, calculus stimulates the appearance of interesting new branches of pure mathematics. In the present century, a few branches of mathematics have developed that do not use calculus. Yet even these are mixed up with subjects related to calculus. Someone studying these branches without a background of calculus would be at a terrible disadvantage; he would meet allusions to calculus; there would be results suggested by theorems in calculus. No person intending to study mathematics seriously could possibly leave calculus out.

Calculus, then, is an indispensable topic, both for the pure and applied mathematician.† And calculus grows from a quite simple idea, the idea of *speed*.

In the past, people often thought of calculus as an extremely difficult subject. Then, particularly in England, teachers began to realize that many things could be done by calculus in a way that was much simpler and more interesting than anything in algebra. In English high schools a student may have two or even three years of calculus. But then some mathematicians say that this is not good; that calculus is really more complicated than it appears, and that it should only be taught by a very well qualified mathematician. Where does the truth lie in all these conflicting views?

A comparison may be helpful. An old lady lives in a quiet village, and every Sunday she drives herself to church. You ask her if it is easy to drive a car. "Oh, yes," she says, "I have no mechanical aptitude, and I find it quite simple." She might find it less simple if she had to drive in the middle of New York, or take a heavy truck across the Rockies. But there is no denying the fact; she can drive a car. And, if she ever did have to drive in heavy traffic, her experience of handling a car would be of some use to her. She would not be so helpless as someone who had never driven at all.

The situation in calculus is somewhat similar. Elementary calculus is like elementary car driving, not difficult to learn and it enables you to do many things you could never manage otherwise. But if you wish to push calculus as far as it will go, you will run into things that are more complicated.

How should calculus be taught then? Should we bother the beginner with warnings that only become important in more advanced work?

---

† A pure mathematician is one who studies mathematics for its own sake. An applied mathematician is one who studies mathematics in order to deal with some aspect of the actual world—science, engineering, medicine, economics, history, etc. Most of the greatest mathematicians of the past were interested in both pure and applied mathematics, and the same is true of some of the best mathematicians living today.

If we do so, the beginner will be confused because he will not see any need for these warnings. If we do not, we shall be denounced by mathematicians for deceiving the young.

I believe the correct approach is to do one thing at a time. When you take a student into a quiet road to drive a car for the first time, he has plenty to do in learning which is the brake and which the accelerator, how to steer, and how to park. You do not discuss with him how to deal with heavy traffic which is not there, nor what he would do if it were winter and the road were covered with ice. But you might very well warn him that such conditions exist, so that he does not overestimate what he knows.

If you try to tell him the whole truth, he probably cannot take it in all at once. An even more important objection is—we do not know the whole truth. Our student is young. Perhaps he will live to drive a car in the first Martian expedition. And who knows what difference in driving technique will be needed on Mars?

Mathematics also is an exploration. As we push out further, we meet new and unexpected situations and we have to revise our ideas. Rules we have used, theorems we have proved turn out to have unforeseen weaknesses. If I were asked to write on a sheet of paper all the statements that I was absolutely sure of, statements that would be true at every time and place, I should leave the paper blank.

In this book I begin with the simple ideas of calculus, with country driving. I do not look for awkward exceptions. In the main, I look at things as mathematicians did in the 17th century when calculus was being developed. I have found that 9th- and 10th-grade students, who are interested in mathematics, can follow this treatment of calculus without difficulty. Towards the end of the book, in the chapter *Intuition and Logic*, I give some examples to show you how things become as you approach the heavier traffic of the big cities. This is to warn you of complexities that can arise. But you should not think of these complexities simply as being *difficulties*. They are not so by any means. Some of the complications are very strange and unexpected and interesting.

Now, the question of what you need to know to read this book; you require the following three things.

(1) *Basic arithmetic*. You must be able to add, subtract, multiply, and divide whole numbers, fractions, and decimals. No knowledge of business arithmetic, percentages, discount, etc. is needed at all. You should have met exponents, and know that, for example, 4^5 is short for $4 \times 4 \times 4 \times 4 \times 4$.

(2) *Basic algebra*. You should know how symbols such as x are used, and be able to add, subtract, multiply, and divide simple algebraic expressions. You should be able to substitute in a formula, for example

to put 3 for x in $x^2 - 1$ and get the answer 8. Negative numbers such as -5 should have been met.

(3) *Graphs.* How to go about drawing a graph should be known. You should have drawn several graphs, and remembered something of what they looked like; for instance, the graphs of $y = x$ and $y = 2x + 1$ are straight lines, while those of $y = x^2$ and $y = 1/x$ are not.

It is particularly important that you have not merely learned algebra as a set of rules, but have some understanding of what algebra is about, how it grows out of arithmetic, and how it is used to say things about arithmetic. A few examples will show what this means.

For instance, the following statements belong to arithmetic:

$$3^2 \text{ is 1 bigger than } 2 \times 4;$$
$$4^2 \text{ is 1 bigger than } 3 \times 5;$$
$$5^2 \text{ is 1 bigger than } 4 \times 6.$$

But these results suggest "the square of any whole number is 1 bigger than the result of multiplying the number before by the number after." For instance, we should guess that 87^2 would be 1 bigger than 86×88. The general result is stated most conveniently in the language of algebra. If n is short for "any number," then "the number before" will be written as $n - 1$ and "the number after" as $n + 1$. Instead of the sentence above, we shall now say, "n^2 is 1 bigger than

$$(n - 1) \cdot (n + 1)\text{"}$$

or, completely in symbols,

$$n^2 = 1 + (n - 1)(n + 1).$$

This equation, which holds for every number n, expresses what we guessed by looking at particular results in arithmetic. Further, it enables us to prove that our guess is correct. By the usual procedures of algebra, we can multiply out and see that the two sides are always the same.

Symbols are thus useful, both for stating what we have guessed, and for proving it correct.

In algebra itself, we often pass from particular results to more general ones. For instance, if you do some algebraic multiplications, such as

$$(x + 3)(x + 4) = x^2 + 7x + 12$$

and

$$(x + 5)(x + 6) = x^2 + 11x + 30,$$

you probably notice something. In the first example, you find on the right-hand side both 7, the sum of 3 and 4, and 12, the product of 3 and 4. In the second example, the same thing happens; 11 is 5 plus 6, while 30 is 5 times 6. We guess that this happens, whatever numbers occur on the left-hand side. In algebraic symbols, we guess that multiplying out $(x + a)(x + b)$ will always give us $a + b$ as the coefficient of x, and ab as the constant term. Our guess, written as an equation, is

$$(x + a)(x + b) = x^2 + (a + b)x + ab.$$

We can now easily prove that our guess is correct. This type of thinking will often be used in this book. We shall collect some evidence. We shall look at it. We shall try to guess some general law.

In doing this, we shall need to observe laws, and to write them in algebraic symbols. For instance, if we are shown the table

x	0	1	2	3	4
y	0	2	4	6	8

,

we easily guess the law that lies behind it. Each number in the bottom row is twice the number that lies above it. The law behind the table is $y = 2x$. In the same way, the law behind the table

x	0	1	2	3	4	5
y	0	1	4	9	16	25

is $y = x^2$. Each number in the bottom row is the square of the number above it.

Incidentally, as a rule, *there is little point in putting a law into words.* It is far easier to see what the formula $y = 3x^2 - 2x + 7$ means, than it is to understand the same formula expressed in words. Algebra is the best language for thinking about laws. Algebra puts the law into a small space. The formula is shorter to write, easier to read, quicker to say, and simpler to understand than the corresponding sentence in ordinary English. If I wanted to be sure that you understood the formula, I would not ask you to turn it into English; I would ask you to work out the table. If you did this correctly, I should know that you understood the instructions contained in the formula.

We cannot always guess straight away the law behind a table. For instance, if I ask you to guess the law behind the table

x	0	1	2	3	4	5
y	0	3	12	27	48	75

,

you may not be able to do this at once. You should be ready to make one or two wrong guesses before you hit on the right one. There is a certain amount of luck in guessing. But if you persist, you should in time hit a trail that leads you home. Here you might, for instance, notice that every number in the bottom row divides exactly by 3. The values of y in fact are

> 3 times 0,
> 3 times 1,
> 3 times 4,
> 3 times 9,
> 3 times 16,
> 3 times 25.

We notice the square numbers 0, 1, 4, 9, 16, 25 now. The law, in fact, is $y = 3x^2$.†

This law, $y = 3x^2$, is in fact one that we shall meet later in this book. We shall need to guess it from a table.

There are systematic procedures for discovering a law to fit a table of numbers,‡ but we shall not go into that here. We shall only be concerned with simple laws, for which plain guessing is good enough.

The Purpose and Limitations of This Book

There are two things that an introductory book should not be; it should not be a cookbook; it should not simply be a collection of theorems with proofs. Both types of book conceal mathematics from the student.

A cookbook is simply a list of rules for solving certain types of problems. The student is expected to learn these rules. But why do these rules work? How were they discovered? What do you do with a problem that does not fit any of the rules?

The theorem-proof-theorem-proof type of book does, in a certain limited sense, explain mathematics to the student. Theorem 1 is at least followed by a proof of Theorem 1, which may throw some light on why Theorem 1 holds. But very much is still left hidden. How did the writer decide that Theorem 1 should come first? How did he decide which

† Note the distinction between $3x^2$ and $(3x)^2$. In $3x^2$ the squaring applies only to x; that is, we take any number for x, we square it, and *then* multiply by 3. Students sometimes take the number for x, multiply by 3, and square the result. This procedure however has the symbol $(3x)^2$.

Thus for $x = 10$, $3x^2$ has the value $3 \cdot (10)^2 = 3 \cdot 100 = 300$, while $(3x)^2$ has the value $(3 \cdot 10)^2 = (30)^2 = 900$.

‡ See *The Mathematics Student Journal*, November 1958 and January 1959. "A Method of Discovery in Algebra."

theorems to include and which to omit? What is the book trying to do? What is the line of thought that lies behind it? How did all these theorems come to be discovered? What should the student do if he wishes to discover further theorems for himself? This last question is perhaps the most important of all. It is a very strange thing that many eminent mathematicians, who think the only thing really worth doing in life is to discover new theorems, often write books which give no hint at all of how a student should try to make his own discoveries.

There are at least four stages in mastering a mathematical result.

(1) You must see clearly and understand what the result states. It is not enough to have memorized certain words. You must know what the result means.

(2) You should collect evidence which shows that it is reasonable that this result should be so; you should feel that this result agrees with your experience of mathematics.

(3) You should know what you can do with the result. It may have applications in science, or it may simply lead to other interesting theorems in pure mathematics. You ought to know what these are.

(4) You should know and understand the formal proof of the result.

I want to make it quite clear that this book does not attempt to provide formal proof of any result whatever. I have not attempted to deal with stage (4) at all. My concern is entirely with stages (1), (2), and (3). I want you to see that the ideas of calculus arise quite naturally, and indeed I want you to discover them for yourself. If we were in a room together, I would confine myself to asking you questions, and you would find that you arrived at calculus by clarifying ideas that you already have in a vague and shadowy form. Between the covers of a book I cannot follow that procedure. But I keep to it as nearly as I can. I am not trying to tell you any particular result. I am trying to call your attention to certain things that you can experiment with for yourself. The evidence that you collect will *suggest* certain conclusions to you. More than that, I do not claim. I do believe, however, that this experience will make it much easier for you when you begin to learn calculus in real earnest. You will have some idea of the direction in which you are traveling.

The first seven chapters differ in some respects from the last three. Chapters 1 through 7 deal with certain topics in some detail. These chapters you may reasonably expect to read and master thoroughly. The last three chapters are much less detailed. They are put there to show you that there is still something to learn when you have mastered the contents of Chapters 1 through 7. Chapters 8 and 9 indicate, in bare outline, some questions that arise in the first year's work on calculus. Chapter 10 raises some more profound questions; it calls your attention to some things that (you would think) could not possibly happen, and yet which do actually happen. Some students find this the most exciting chapter in the book.

Chapters 8, 9, and 10, then, are something in the nature of a preview, a sample of things to come. Their aim is more to indicate the kind of question that lies ahead than to give you a thorough account. So do not be surprised if these chapters leave some unanswered questions in your mind.

After Chapter 10 you will find a "Guide to Further Study." This begins with quite elementary texts on calculus and goes on to quite advanced topics. The later parts of this "Guide" will mainly be of interest to the rather exceptional student, who reads this little book while in the 9th grade, and works steadily at calculus for the remainder of his high-school career. Only a few students are capable of doing this, but those who can do it should be given every encouragement to go ahead.

The book concludes with a list of technical terms. This list was compiled after the rest of the book had been written. So far as understanding the book is concerned, this list could have been omitted. It was felt, however, that readers might like to know the official names of the ideas they had met, and also that this would be helpful in reading more formal books on calculus.

CHAPTER TWO

The Study of Speed

We are going to investigate speed, the speed of a moving object. How can we see clearly what a moving object is doing? We might make a "movie" of an object moving along a straight line. Suppose we have a camera that makes a picture every tenth of a second. Suppose successive pictures are as shown in Fig. 1. What is the little object doing? Every tenth of a second, it moves up 1 inch. It seems to be moving with a steady speed of 10 inches a second.

On another occasion, we might obtain the pictures shown in Fig. 2. Here, the object advances 2 inches between each picture and the next. It has a steady speed of 20 inches a second.

Let us look at something with a varying speed. Suppose an object is accelerating. Between the first and second pictures it might cover 1

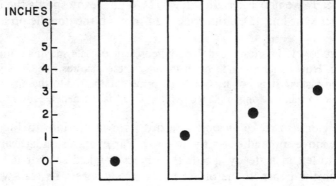

Figure 1

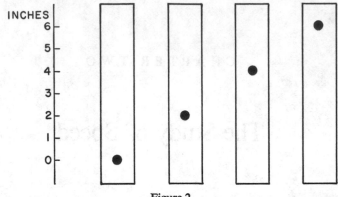

Figure 2

inch; between the second and third, 2 inches; between the third and
fourth pictures, 3 inches. Its record would be as shown in Fig. 3.

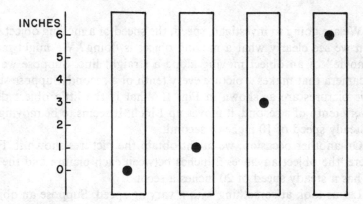

Figure 3

Already we notice certain things. (1) With steady speeds the dots lie
on a straight line, (2) with accelerated motion, the dots lie on a curve.

QUESTION 1. Figures 1 and 2 both represent objects moving with steady
speeds. How could one tell, by examining these pictures, which object was
moving faster? It is not necessary to bring numbers into the answer. It is
possible to tell, at a single glance, which object is the faster. How?† _p. 117_

We can also make an object record its own motion. In Fig. 4, the
object moves up and down the line *PQ*. Paper passes underneath from
right to left at a steady speed; the object is inked so that it leaves a
trail on the paper. If the object has a steady speed, its trail will be a
straight line.

† Answers to problems will be found at the back of the book.

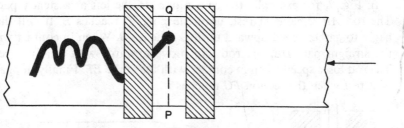

Figure 4

QUESTION 2. Fit the records shown in Fig. 5 to the descriptions:

(a) Moving up rapidly.
(b) Moving up slowly.
(c) Stationary.
(d) Moving down slowly.
(e) Moving down fast.

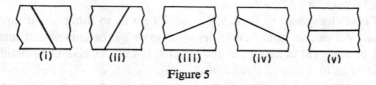

Figure 5

QUESTION 3. Fit the records shown in Fig. 6 to the descriptions:

(f) Starting from rest and gradually gaining speed upwards.
(g) Rising fast at first and gradually slowing down to rest.
(h) Starting from rest and gradually acquiring speed downwards.
(i) Falling fast at first and gradually being brought to rest.

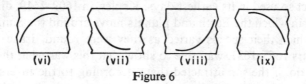

Figure 6

No special equipment is needed, if you want to demonstrate the connection between curves and movement. The simplest thing is to draw the curve first, and then pass it behind a narrow slit; the arrangement is similar to that of Fig. 4. You will only be able to see a small part of the curve through the slit, and this will give you the impression of a point rising and falling.

This has an engineering application. If we want to make an object behave in a particular way, we can do so by means of a suitably shaped cam.

In Fig. 7, for example, the cam moves to the left at a steady pace. The rod AB remains at rest, until the point C reaches B. It will then begin to gather speed upward until D reaches B. When in contact with the straight part DE, the rod will move upwards with steady speed. The rod loses speed when in contact with the curve EF. Finally, it again is at rest when the section FG reaches it.

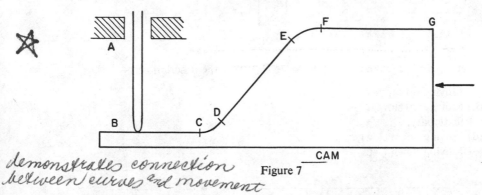

Figure 7

demonstrates connection between curves and movement

Curves like those in Figs. 5, 6, and 7 help us to think about movement. We can see the curves; details appear in the curves that might not be apparent in the actual movement; the curves give us something definite to look at and think about.

The work we have done also tells us something about the scope of calculus. Calculus begins as the study of speed. But in thinking about speed, we have been led to the curves drawn above. These curves could be described in terms of speed. For example, curve (viii) could be described as the curve that records the movement, when an object moves upward faster and faster. So calculus can be used not only to describe *movement* but also to describe the *shapes of curves*. Calculus was in fact so used in its earliest days. Kepler, in 1609–1619, discovered the paths in which the Earth and planets move around the Sun, and the way in which their speeds varied as they went round. Isaac Newton, in the years 1665–1687, was able to show that this was what the planets ought to do, if the sun attracted them according to the inverse-square law. Thus, with the help of calculus, he accounted for both the speeds and the curves. It impressed men very much that the complicated behavior of the solar system could be deduced from three or four very simple assumptions—Newton's laws of motion and his law of gravity. Newton's laws, and his application of calculus to astronomy, have a renewed interest today, when not only can we look at the planet Mars but some of us may be able actually to go there. Calculus would be used to calculate the possible orbits from the Earth to Mars, and to decide which orbit would require the least fuel.

Calculating Velocity

Now let us turn to some simple calculation. How do we work out the velocity of an object? Suppose, for example, a car is traveling along a straight road, a turnpike say. At 2 o'clock the mileage recorder shows 70 miles. At 5 o'clock, the mileage is 220 miles. *Suppose the car has been traveling all the time at a steady speed* (this is most unlikely in practice!). How fast has it been going? This is not a difficult question. Subtracting 70 from 220, we see that the car has gone 150 miles. Subtracting 2 from 5, we see that it has taken 3 hours to do this. We divide 150 by 3 and get 50. So the speed is 50 mph.

Our reason for doing this simple piece of arithmetic is to study the *method*, rather than the answer. We want to extract from it a formula for velocity. We bring some symbols in. Let s miles be the reading of the mileage recorder at the time t hours. Thus, $t = 2$ would indicate that the time was 2 o'clock, and $s - 70$ would indicate that the car had gone a total distance of 70 miles. The information we had in the question above could be put in a table like this:

t	2	5
s	70	220

But we want to get away from the particular numbers 2, 5, 70, 220. We want a formula for giving the velocity between any two times and any two places. So we bring in some more symbols.

Generalized problem. "At a hours, the mileage is p miles. At b hours, the mileage is q miles. The car moves at a steady speed. Find its velocity, v miles an hour."

We do the same steps as we did in the particular arithmetical problem, but we replace the particular numbers by the corresponding symbols. a should appear now, where 2 appeared in the arithmetic; b replaces 5, p replaces 70, q replaces 220. The table is:

t	a	b
s	p	q

In the arithmetic, we began by subtracting 70 from 220. In the algebra, we subtract p from q. So the car has gone $(q - p)$ miles. How long has it taken to do this? Instead of subtracting 2 from 5, we subtract a from b. The car has taken $(b - a)$ hours. <u>To find the velocity, we divide the number of miles gone by the number of hours taken.</u> This gives us

Formula (1)

(constant velocity)

$$v = \frac{q - p}{b - a}.$$

It is most important to remember that this formula holds only if the car has a steady speed—if it moves at a constant velocity.

Suppose, for example, a car driver drove 30 miles in one hour, then spent 3 hours having dinner, suddenly realized how late it was, drove for an hour at 95 mph, and then had an accident. It would be no good for this driver to say, "I have been out for 5 hours and have covered 125 miles. So my speed can only have been 25 mph. The accident was not my fault." At the moment of the accident, his speedometer was showing 95 mph. That is what we mean by velocity; what the speedometer shows *at a particular instant*. It has nothing to do with ancient history. Maybe this driver had not used his car for a year. Then he could say that he had only covered 125 miles in a year, which is 0.014 miles an hour. Everyone would call this a ridiculous defense. I only emphasize this point because many students of calculus behave exactly like this man. They remember formula (1). It is so simple that they use it even in situations where it gives the most ridiculous results.

Formula (1) works only when an object travels with constant velocity. If the velocity varies a little, then formula (1) gives us, not the exact velocity, but a reasonably close estimate of it. For example, the speed of a car does not vary much in one second. Formula (1) would give *a reasonable estimate* of a car's speed, if one observed the distance the car went in a second. Such evidence might be available if someone had been taking a movie when a car crashed, and it would be quite reasonable to produce that movie in a law court. In calculus, we use something of the same procedure. We are mainly interested in cases where the velocity is varying all the time. So we cannot simply quote formula (1). That would be quite wrong. What we do, is to use formula (1) to *estimate* the velocity; by using shorter and shorter times, we try to arrive at some conclusion.

Negative Velocity

One curious result can be drawn from formula (1) even in the case of steady velocity. Suppose the car is going *backwards*. This happens rarely or never with cars, so our example is somewhat unreal. However, in science the situation frequently occurs; for example, a stone, thrown straight up into the air, rises for a certain time, and then falls. When falling, it is returning to its original position, like a car backing. Suppose then, a car capable of driving backwards at a steady speed for two or three hours. How would its table look? Something like this—

t	3	5
s	80	60

At 3 o'clock, it would be 80 miles from home; at 5 o'clock, only 60 miles. In 2 hours, it has returned 20 miles; evidently, it has been backing at 10 mph.

What does formula (1) give? We have to put

$$a = 3, \quad b = 5, \quad p = 80, \quad q = 60.$$

This gives

$$v = \frac{q - p}{b - a} = \frac{60 - 80}{5 - 3} = \frac{-20}{2} = -10.$$

We know the car is backing at 10 mph. The formula gives $v = -10$.

There are two ways of dealing with this situation.

(1) We might say, "It is absurd to have negative velocities. A velocity cannot be less than zero. If a car is going backwards, you must use a different formula. Formula (1) just does not apply then."

(2) We might say, "We will use formula (1) always when something moves with a steady speed. If formula (1) gives us a negative answer, we shall know that the object is moving backwards."

Policy (2) has been found to be much the most convenient. If we used policy (1), it would double our work; we should have one set of rules for things that are rising, another set for things that are falling. Policy (2) allows us to have a single formula. If, at the end, the answer comes out negative, we know what that means. Usually, in a car, the speedometer shows only speeds *forward*. What we are doing now is rather more like what happens on a ship, where you have "full speed ahead" and "full speed astern." One could imagine a car with an extended speedometer, that went past zero to show "−5 mph" when the car was backing at 5 mph, "−10 mph" when it was backing at 10 mph, and so on.

In physics, the word *velocity* is commonly used when *direction* is being taken into account; *speed* is used when you are simply concerned with how fast an object is moving, and not bothering whether it is moving forwards or backwards. Thus a car advancing at 10 mph has a *velocity* of +10 mph; when backing at 10 mph, it has a *velocity* of −10 mph. In both these cases, the *speed* is 10 mph. This distinction will not play any part in this book. We shall always be concerned with velocity. For example, we might record various movements as in Fig. 8.

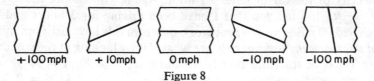

+100 mph   + 10 mph   0 mph   −10 mph   −100 mph

Figure 8

Rates of Change

If we are traveling in a car, the velocity of the car is the rate at which the mileage increases. Velocity is the rate of change of distance gone. Calculus is concerned with how fast things change. The thing changing need not be a distance. We may ask, "How fast is that man growing rich?" "How fast is this car's tank being filled with gas?" These are rates of change—the rate of change of a bank account; the rate of change of the amount of gas in the tank.

It is convenient to have a symbol for "the rate of change of." We shall use a very simple one, the symbol

$$'.$$

If f measures any quantity, f' measures the rate at which that quantity is growing (f' is read "f prime" or "f dashed").

For example, if a boy is h inches in height when he is n years old, h' means the rate at which he is growing, in inches a year.

If a car goes s miles in t hours, s' means the rate, in mph, at which the mileage grows. s' miles an hour is in fact the velocity of the car.

If there are g gallons of gas in a tank after t seconds of filling, g' means the rate at which gas is entering the tank, measured in gallons a second.

If a man has m dollars when he is n years old, m' is the rate at which his wealth is increasing, in dollars a year.

Note here the distinction we made earlier: m' is not the same as m/n. If a man has \$3000 when he is 30 years old, it does not in the least follow that his wealth is increasing at the rate of \$100 a year. You could only draw this conclusion if you knew that, from the time he was born, he had been saving money at a steady rate. It might be that he had nothing at all until he was 27, and in the last three years he has been saving steadily at \$1000 a year. In that case, m' would be 1000. On the other hand, it may be that he is having a difficult time now, and is actually losing money at \$500 a year. In that case $m' = -500$. m' has nothing to do with ancient history. It measures what is happening now.

If s miles is the distance a car has gone in t hours, s' denotes the velocity of the car in miles an hour. Again, you cannot assume that $s' = s/t$. If I tell you that I have been driving for 3 hours and have covered 90 miles, you cannot work out from this how fast I am moving at this moment. You can only see what s' is by looking at the speedometer. I may be traveling at sixty. In this case, $s = 90$, $t = 3$, $s' = 60$. Or my car may be at rest. In that case $s = 90$, $t = 3$, $s' = 0$. I may

S = distance covered (in miles)

S' = velocity; rate of change (in mph)
at a particular
moment - INSTANTANEOUS

AMOUNT

RATE OF
CHANGE —
involves
TIME

even be backing at 10 miles an hour. Then $s = 90$, $t = 3$, $s' = -10$. ↓

All this merely amounts to saying that, if I tell you what time it is and where I am, you cannot tell me how fast I am moving. However it is necessary to emphasize this. Students seem to have had drilled into them "velocity is distance divided by time." This is so *only in the case of steady velocity*. But the whole point of calculus is to study variable velocity, as when a ball is falling to the earth or a rocket taking off from the earth.

s' then is the number to which the speedometer is pointing at any particular moment.

EXAMPLES. Translate into calculus symbolism:

(1) After I had been traveling for 5 hours, I had covered 120 miles and was driving at 40 mph. ← *instantaneous velocity*

ANSWER. For $t = 5$, $s = 120$ and $s' = 40$. *instantaneous velocity*

(2) After 2 hours' driving, my speedometer showed 50 mph and after 3 hours it showed 45 mph.

ANSWER. For $t = 2$, $s' = 50$. For $t = 3$, $s' = 45$.

(3) For the first two hours, I drove at a steady speed of 40 mph.

ANSWER. $s' = 40$ for every value of t from 0 to 2. ✳

Finding Velocity in Simple Cases *constant velocity*

There are some cases in which velocity can be found by arithmetic alone. These cases are, of course, not very interesting or exciting; the interesting results come in the problems where new methods are needed. These simple cases, however, can get us used to the s' symbolism.

Suppose the mileage on my car is zero, and I drive at a steady velocity of 10 mph for a certain time. The table giving my mileage at any time is

t	0	1	2	3	4
s	0	10	20	30	40

Here, $s = 10t$ is the law. What is s'? We said at the outset that my velocity was steady at 10 mph, and s' measures my velocity. So $s' = 10$. Let us set this out formally.

Result A. If

$$s = 10t,$$

$$s' = 10.$$

S = distance
t = time
S' = rate at which velocity is changing

Since my velocity is 10 mph all the time, $s' = 10$ does not simply mean that s' is 10 at some particular instant, but that *at any instant* during the motion s' has the value 10. $s = 10t$ is a law for the motion in the sense that it tells you where the car is at any time. If you ask, "Where is the car after $1\frac{1}{2}$ hours?" I substitute $t = 1\frac{1}{2}$ in the formula $s = 10t$ and get $s = 15$. $s' = 10$ is also a law, in the sense that it tells me the velocity at any time; it says that the velocity is always 10.

Here we have an example of <u>one of the first problems of calculus: given a law that tells you where an object is at any time, find a law for its velocity at any time.</u>

constant velocity : **Exercises**

1. To begin with, the mileage of my car is zero. I drive at a steady velocity of 20 mph. What law gives my position at any time? What is my velocity at any time? Write the answers to both questions as equations.

2. The position of a car at any time is given by the equation $s = 30t$. What is the mileage when $t = 0$? when $t = 1$? when $t = 2$? when $t = 3$? What is the velocity of the car? What equation gives s'?

3. The position of a car at any time is given by the equation $s = 40t$. Find the equation for the velocity of the car.

4. Complete the statement, "if $s = 50t$, $s' = \ldots$".

5. If k stands for any fixed number (like 20, 30, 40, 50 in the preceding examples) and $s = kt$, then $s' = \ldots$?

In the examples just considered, we started each time with zero mileage. This however is not necessary. Consider the law $s = 10t + 3$. The table for this is

t	0	1	2	3	4
s	3	13	23	33	43

Here, the mileage recorder showed 3 at the beginning. The table shows that the car covers 10 miles with every hour that passes. The velocity is 10 mph, and so $s' = 10$. We thus have

Result B. If

$$s = 10t + 3,$$
$$s' = 10.$$

Exercises

Investigate in the same way the velocity s' corresponding to the laws (1) $s = 10t + 5$, (2) $s = 10t + 7$, (3) $s = 10t + 9$.

If $s = 10t + c$, where c is any fixed number, what is s'? What is the value of s' for (4) $s = 20t + 3$, (5) $s = 20t + 5$, (6) $s = 20t + 7$, (7) $s = 20t + 9$?

Can you draw any conclusion from the above examples? Can you write down straight away the velocity s' corresponding to the laws (8) $s = 30t + 7$, (9) $s = 50t + 9$, (10) $s = 40t + 23$, (11) $s = 30t + 20$, (12) $s = 50t + 150$?

If you drew illustrations of the motions considered in these examples, such as we had in Figs. 1 through 8, what would these illustrations look like?

CHAPTER THREE

The Simplest Case of Varying Speed

Velocity at an Instant

Steady velocity is too simple to be very exciting. We now turn to the real problem, the question of variable velocity.

It should be emphasized that the quantity v or s', for which we are seeking, is intended to measure velocity *at an instant*. In everyday life we find this quite simple; we glance at the speedometer of a car; the needle points to 60 mph and we conclude that 60 mph is our speed at this instant. But when we start to examine what this means, we meet a certain paradox. The very idea of velocity seems to involve *two* times, the beginning and end of an interval. We measure velocity in miles an hour, and these words imply that we see how far an object goes in a certain time. If the time allowed is zero, the distance the object goes is zero. However fast it may be going, two photographs of it taken at the same time will show it at the same place.

If in formula (1) we were to try to discover the velocity at an instant, by making a and b coincide, then p and q also would coincide, and the formula would give us $0 \div 0$ as the velocity—which does not help us at all.

We have used curves to record the movement of objects. A steep line corresponds to an object moving fast; a gentle slope to an object moving slowly (Figs. 5 and 6). So our question could be posed in terms of curves. Instead of saying, "What is the velocity at this instant?" we could ask, "What is the steepness of the curve at the point P?" (see Fig. 9). This seems a sensible sort of question. We would agree, for

Figure 9   Figure 10

example, that, for the curve shown in Fig. 10, the steepness at the point R is greater than at the point Q. We know what we mean when we say this. But suppose the curve were covered up in such a way that we could only see the point Q itself (Fig. 11). We should have no idea how steep the curve was at Q. Suppose the screens are moved a little apart, so that we see just a little bit of the curve near Q (Fig. 12).

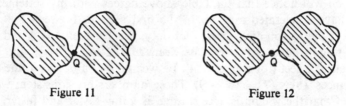

Figure 11   Figure 12

Now we can see what the steepness is at Q; it does not matter how little of the curve is exposed, so long as we can see a piece of curve on each side of Q.

Accelerated Motion

Let us now take a particular case of motion with variable velocity, and see how the velocity at any instant can be calculated. This example that we are going to study is in fact of importance in physics; it is the type of motion usually studied at the beginning of a course in mechanics. It could be produced by the apparatus shown in Fig. 13.

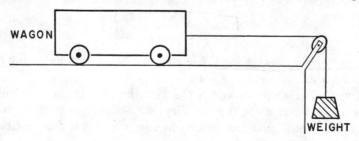

Figure 13

If the wagon weighed 15 ounces, the weight would have to be somewhat more than 1 ounce. "Somewhat more" because there would be friction acting at the wheels of the wagon; by adjusting the weight, the desired motion could be obtained, namely, that given by the table

$$t \quad 0 \quad 1 \quad 2 \quad 3 \quad 4$$
$$s \quad 0 \quad 1 \quad 4 \quad 9 \quad 16 \quad .$$

It is understood that s feet is the distance gone by the wagon in t seconds. The table of course fits the law

$$s = t^2.$$

You will notice that the table above agrees with my statement that we have accelerated motion. In the first second, between $t = 0$ and $t = 1$, the wagon advances 1 foot only. But between $t = 1$ and $t = 2$, the wagon advances 3 feet. Between $t = 2$ and $t = 3$, the wagon advances 5 feet (for $5 = 9 - 4$). Between $t = 3$ and $t = 4$, the wagon advances 7 feet ($7 = 16 - 9$). These numbers are consistent with the belief that the wagon is accelerating, is going faster and faster, as the weight pulls it forward.

Suppose now we try to estimate the velocity at the instant when $t = 3$. In the second before this instant, from $t = 2$ to $t = 3$, the wagon covers 5 feet. In the second after this instant, from $t = 3$ to $t = 4$, the wagon covers 7 feet. It seems reasonable to guess that the velocity at the instant $t = 3$ lies between 5 and 7 feet a second.

Students nearly always ask, "Couldn't we take the average of 5 and 7, and say that the velocity is 6 feet a second?" Unfortunately, this answer is correct for this particular example. I say, "Unfortunately," because, as a rule, taking the average does not give the correct velocity. In fact it hardly ever gives the correct velocity. Only when the law is of the type

$$s = at^2 + bt + c$$

does taking the average work. We shall see below that averaging gives a wrong result for the law $s = t^3$.

If you will take my word for this, for the time being, we shall set aside the guess that the true velocity is exactly halfway between our estimates 5 and 7, and merely use our conclusion that the velocity lies *somewhere* between 5 and 7.

How can we narrow down this margin? We agreed earlier that the shorter the time interval was, the better estimate one should get for the velocity. It seems a good idea to take a shorter interval. Instead of one second before and after $t = 3$, we try half a second before and after. By substituting in the formula $s = t^2$, we obtain the little table

$$t \quad 2\tfrac{1}{2} \quad 3 \quad 3\tfrac{1}{2}$$
$$s \quad 6\tfrac{1}{4} \quad 9 \quad 12\tfrac{1}{4}$$

What use has the wagon made of these half seconds? In the half second between $t = 2\tfrac{1}{2}$ and $t = 3$, s has grown from $6\tfrac{1}{4}$ to 9. That is, the wagon has covered $2\tfrac{3}{4}$ feet. Two and three-quarters feet in $\tfrac{1}{2}$ second suggests a velocity of $2\tfrac{3}{4} \div \tfrac{1}{2}$, which equals $5\tfrac{1}{2}$ feet a second.

In the half second after $t = 3$, the wagon covers $12\tfrac{1}{4} - 9$, that is, $3\tfrac{1}{4}$ feet. Three and one-quarter feet in $\tfrac{1}{2}$ second suggests the velocity $3\tfrac{1}{4} \div \tfrac{1}{2}$, that is, $6\tfrac{1}{2}$ feet a second.

So we now think the velocity should lie between $5\tfrac{1}{2}$ and $6\tfrac{1}{2}$ feet a second.

But why stop at a half? Why not go to shorter and shorter intervals, getting better and better estimates?

If we take one-tenth of a second before and after $t = 3$, we get the little table

$$t \quad 2.9 \quad 3 \quad 3.1$$
$$s \quad 8.41 \quad 9 \quad 9.61$$

by means of the formula $s = t^2$. In the tenth of a second before $t = 3$, the wagon advances 0.59 feet; this suggests a velocity of $0.59 \div 0.1 = 5.9$ feet a second. In the tenth of a second after $t = 3$, the wagon advances 0.61 feet, which suggests a velocity of $0.61 \div 0.1 = 6.1$ feet a second. We now think the velocity should lie between 5.9 and 6.1 feet a second.

By exactly the same method, if we take one-hundredth of a second before and after $t = 3$, we are led to believe that the velocity lies between 5.99 and 6.01 feet a second. By taking one-thousandth of a second, we are led to believe the velocity is between 5.999 and 6.001 feet a second.

We collect these results in the form of a table:

By considering intervals of	We are led to believe that v, the velocity in feet per second, lies between		
1 second	5	and	7
0.1 second	5.9	and	6.1
0.01 second	5.99	and	6.01
0.001 second	5.999	and	6.001

Our last estimate here, using 0.001 second, pins v down to a very narrow region, since 5.999 and 6.001 differ by only 0.002. But of course there is no need to stop at an interval of 0.001 second. We could use a millionth or a billionth of a second, and get even more accurate estimates of v. In fact, there seems to be no limit to how accurately we can estimate v. For the table above shows a very marked pattern. I should imagine you can guess how the table would continue.

Exercise

Without making any calculations, guess the estimates of v that would correspond to intervals of 0.0001 second and 0.00001 second. Check your guesses by actual calculation.

I imagine you had no difficulty in seeing how the table would continue. Each row we go down, we find one more 9 in 5.99 ... 9 and one more zero in 6.00 ... 01. The estimates are coming closer and closer together. Any particular estimate leaves some uncertainty about the value of v, even though this uncertainty may be very small. But if we take *all* the estimates into account, this uncertainty disappears. There is only one number that is bigger than 5.999 ... 9, however many nines are written, and smaller than 6.000 ... 01, however many zeros are written. That number is 6.

So, although we spoke of estimating the velocity, and an estimate usually implies some degree of error or uncertainty, yet there is no uncertainty at all in our final answer. 6 is the only number that satisfies all the estimates, as they close in from the right and the left.

All this arithmetic thus leads us to the conclusion that, if a body moves according to the law $s = t^2$, when $t = 3$ its velocity is given by $v = 6$. ✶

The purpose of this explanation is that you should now be able to work out for yourself the values of v corresponding to $t = 1$, $t = 2$, $t = 4$, and $t = 5$. When you look at your answers you should notice a certain law.

I must make sure that you understand the method for finding v corresponding to any given value of t. In classes, some students see the point of the method straight away; but there are always some who have to have it explained more than once. So, for readers who need it, I will indicate how to get clear about the method. It is important that you should understand this method, for the next stage of the work requires you to discover the first result of calculus; you will feel much happier and more confident if you discover it for yourself, than if I have to tell you.

✶ ONLY when $T = 3$ does $V = 6$ (what other values at other Ts)

First of all, you must be clear as to the idea behind the method. Formula (1), which is often spoken in the form "velocity is distance divided by time," applies only to <u>constant</u> velocities. When the velocity is <u>varying</u>, distance divided by time gives the *average velocity* only; the <u>actual</u> velocity <u>at any instant</u> may be more or less than the average velocity. However, we consider shorter and shorter intervals of time; we hope that this gives less and less opportunity for the velocity to vary, so that <u>the average velocity</u>, over a <u>very short</u> interval, should be <u>a good estimate</u> of the true velocity.

Second, you need to be able to carry through the actual calculation. If you find difficulty in organizing the work, you may find it helpful to adapt the argument of pages 24–25; go through the same kind of steps, but work out the velocity for $t = 2$ instead of $t = 3$. Then go through the steps again, but this time find v for $t = 4$. Of course, do not be content just to go through the arithmetic. Think all the time what you are doing and why that should be done.

When you have worked out v corresponding to $t = 1$, $t = 2$, $t = 4$ and $t = 5$, complete the following table:

$$t \quad 1 \quad 2 \quad 3 \quad 4 \quad 5$$
$$v \quad \ldots\ldots \quad 6 \quad \ldots\ldots \quad .$$

After completing the table, you should notice a law connecting v and t. The law is $v = \ldots\ldots$

It is best if you do not read further until you have successfully completed this work.

* * *

The Law for the Velocity

If you carry through the arithmetic correctly, you should arrive at the following result:

$$t \quad 1 \quad 2 \quad 3 \quad 4 \quad 5$$
$$v \quad 2 \quad 4 \quad 6 \quad 8 \quad 10 \quad .$$

Each number in the second row is exactly twice the number above it. So the law is $v = 2t$. If we use the sign ' introduced on page 18, we may use s' instead of v. We then have a new result to put beside our results A and B on pages 19, 20.

Result C. If

$$s = t^2,$$
$$s' = 2t.$$

This very simple result has come out of long calculations. We shall in a moment consider how it might be found more shortly. However, these long calculations have in no way been wasted. This work should have given you a feeling for what is happening. Many students read short proofs of this result in calculus texts; they pass quickly over the algebra, and they never realize what it really means.

Let us now see how algebra could be used to reduce our work, and also to make the argument more convincing.

When we considered how far the object went according to the law $s = t^2$ between the times $t = 2.99$ and 3, we had the messy arithmetical task of finding the square of 2.99. This work can be simplified by algebra. There is a standard result of algebra

Formula (2) $(a + b)^2 = a^2 + 2ab + b^2.$

If we put $a = 3$, $b = -0.01$, we get $a + b = 2.99$. So formula (2) gives us

$$(2.99)^2 = 3^2 + 2 \cdot 3 \cdot (-0.01) + (0.01)^2$$
$$= 9 - \quad 0.06 \quad + 0.0001$$
$$= 8.9401.$$

This method involves less work, and is less likely to lead to a mistake than the usual method of elementary arithmetic.

However, we can make greater use of algebra than simply to shorten the calculations. On page 25, we observed a column containing the numbers 5; 5.9; 5.99; 5.999; and we made a guess as to how this column would continue. By using algebraic symbols, we can avoid this guess. Instead of considering, one at a time, the intervals

$$\text{between} \quad 3 \quad \text{and} \quad 3 + 0.1;$$
$$\text{between} \quad 3 \quad \text{and} \quad 3 - 0.1;$$
$$\text{between} \quad 3 \quad \text{and} \quad 3 + 0.01;$$
$$\text{between} \quad 3 \quad \text{and} \quad 3 - 0.01; \quad \text{etc.,}$$

we can notice that all these intervals are particular cases of the interval

$$\text{between} \quad 3 \quad \text{and} \quad 3 + h.$$

The particular cases can be got by substituting 0.1; -0.1; 0.01; -0.01; respectively for h. Since we can equally well substitute 0.0000001 or 0.000000001 for h, we are thus enabled to deal with the intervals of one-millionth or one-billionth, or any other number for that matter, all at one blow, by an algebraic calculation.

We now carry this idea into practice. We want to find the velocity at $t = 3$. So we consider a short interval, from $t = 3$ to $t = 3 + h$. We must find where the object is at these times. The position is determined by the formula $s = t^2$. When $t = 3$, $s = 9$. When $t = 3 + h$, $s = (3 + h)^2 = 9 + 6h + h^2$. So we have the table

$$
\begin{array}{ccc}
t & 3 & 3 + h \\
s & 9 & 9 + 6h + h^2
\end{array}
$$

We now use "distance divided by time" to estimate the velocity. How far has the object gone during this interval of time? We take the difference between the numbers in the row giving s. The object has gone $6h + h^2$ feet during the interval. How long is the interval of time? We take the difference between the numbers for t. The interval is of length h seconds. Division gives us our estimate of v, namely,

$$\frac{6h + h^2}{h}.$$

distance/time = average velocity

The expression can be simplified. Since

$$6h + h^2 = h \cdot (6 + h),$$

on dividing both sides by h we have

$$\frac{6h + h^2}{h} = 6 + h.$$

Exercise

In the above expression substitute in turn, the values 1; 0.1; 0.01; −1; −0.1; −0.01 for h, and check that the results agree with numbers in the table on page 25, the positive values of h giving one column and the negative values of h the other.

When h is positive, we are considering a little interval just after $t = 3$. Our estimate of v is then $6 + h$, just a little more than 6.

When h is negative, we are considering a little interval just before $t = 3$. Our estimate of v is then just a little less than 6. (For example, if $h = -0.01$, the estimate $6 + h$ is $6 + (-0.01)$, that is 5.99, just less than 6.)

The shorter we make the interval, the closer our estimate comes to 6. We are thus led to the conclusion that $v = 6$.

We have just found, by algebra, the velocity corresponding to $t = 3$. Now there is nothing special about the number 3. The work would have been just as easy for any other number. Here again is the sort of situation where algebra can help; we can use a symbol for "any number" and find v corresponding to any value of t.

Suppose then, we try to find the velocity when $t = a$, where a stands for "any number." The work will follow exactly the same plan as it did for $t = 3$. We can go through this work, step by step, but writing a wherever 3 came before.

Exercise

Do this, if you can, before reading it below.

*   *   *

We shall have the table

t	a	$a + h$
s	a^2	$a^2 + 2ah + h^2$

Distance gone during the interval is found from the difference between the two numbers in the row for s; the distance is $2ah + h^2$ feet. The time taken is found from the difference between the numbers in the row for t. The time taken is h seconds. Division gives the estimate

$$\frac{2ah + h^2}{h}$$

for v, and this simplifies to

$$2a + h.$$

Now h is a very small number; it represents the length in seconds of the interval during which we observe the motion; the shorter this interval, the better the estimate of v. As h gets smaller and smaller, $2a + h$ approaches $2a$. We conclude

$$v = 2a.$$

Thus, if $t = a$, $v = 2a$. In words, "If t is any number, v is twice that number." This confirms the guess we made on page 27. But there our evidence was limited to the numbers 1, 2, 3, 4, 5. By using algebra, we see that $v = 2t$ holds for *every* value of t. Of course a need not be a whole number; the laws of algebra hold equally well for fractions and irrational numbers.

You may wonder why we bother to bring the number a into the discussion. Why write $t = a$, $v = 2a$ instead of just $v = 2t$? The reason is that, at the beginning of our work we had to consider an interval of h seconds, from $t = a$ to $t = a + h$. If we had tried to do without a, we might have found ourselves talking about the interval from $t = t$ to $t = t + h$, which sounds somewhat peculiar!

A Useful Symbolism

In discussing motion, we continually use phrases such as "where the body is at a certain time," or the corresponding algebraic phrase, "the value of s corresponding to a particular value of t." Seeing this phrase is used so often, it is convenient to have an abbreviation for it. We shall use $s(a)$ to stand for "the value of s corresponding to $t = a$." Thus, in the table

$$t \quad 0 \quad 1 \quad 2 \quad 3$$
$$s \quad 0 \quad 1 \quad 4 \quad 9$$

9 is the value of s corresponding to $t = 3$; we can save a lot of space by expressing this in the abbreviated form $s(3) = 9$. For the same table, $s(0) = 0$, $s(1) = 1$, $s(2) = 4$.

When we are discussing velocities, we consider the interval of time from $t = a$ to $t = a + h$. We then examine where the object is at the beginning and end of this interval. Its position is specified by the value of s. The value of s corresponding to $t = a$ can now be indicated by $s(a)$, and the value of s corresponding to $t = a + h$ by $s(a + h)$.

Thus, in this interval, the object covers a distance of $s(a + h) - s(a)$ feet. The time taken is $(a + h) - a = h$ seconds. Thus the average velocity during the interval is

Formula (3)
$$\frac{s(a + h) - s(a)}{h}$$

feet per second.

Procedure for Determining Velocity

We are now able to describe the steps by which we found the law for velocity in our work above. The purpose of describing the procedure is, of course, so that we can apply it to other laws besides $s = t^2$.

(1) We began with a law giving s in terms of t.

(2) We then considered the average velocity during the interval between $t = a$ and $t = a + h$. This led us to the expression given in formula (3) above, namely,

$$\frac{s(a + h) - s(a)}{h}.$$

(3) We allowed h to become smaller and smaller. Thus h approached the value zero. We then found that

$$\frac{s(a + h) - s(a)}{h}$$

approached a certain value.

(4) That value we regarded as giving the velocity at the instant $t = a$.

In our symbolism, this result would be written $v(a)$ or $s'(a)$, for it gives the value of v or s' at $t = a$.

... udy of the law
... ciple is involved
... power. I hope
... s is suggested in

... 5 5.001

... ecimals only. From
... 3, 4, 5. (*Hint*. Each
... at law would give
... e 8.) *Conclusion:* if

... ally by considering
... $- h$ (adapt the argu-

... law for the velocity

4. The velocity law for $s = $ ... was found in ... you solved questions
1, 2, and 3, you found the velocity laws for $s = t^3$ and $s = t^4$. Examine
these results. Do they enable you to guess the law for $s = t^5$? for $s = t^6$?
for $s = t^n$, where n is any whole number? Does your guess give the correct
result for $n = 1$?

The Law $s = t^n$

Exercise 1, I hope, led you to the conclusion that, if $s = t^3$, then $s' = 3t^2$. Exercise 2 should lead to the same conclusion by an algebraic argument. For the little table will read

$$
\begin{array}{lll}
t & a & a+h \\
s & a^3 & a^3 + 3a^2h + 3ah^2 + h^3
\end{array} \quad .
$$

Thus the estimate for v is

$$
\frac{3a^2h + 3ah^2 + h^3}{h} = 3a^2 + 3ah + h^2.
$$

As h approaches zero, both $3ah$ and h^2 approach zero. So the expression above approaches $3a^2$. Thus, when $t = a$, $v = 3a^2$, as we hoped.

In Exercise 3, the algebraic argument involves much less work than the arithmetical. It leads to the little table for $s = t^4$,

$$
\begin{array}{lll}
t & a & a+h \\
s & a^4 & a^4 + 4a^3h + 6a^2h^2 + 4ah^3 + h^4
\end{array} \quad ,
$$

and hence to the estimate for v,

$$
\frac{4a^3h + 6a^2h^2 + 4ah^3 + h^4}{h} = 4a^3 + 6a^2h + 4ah^2 + h^3.
$$

When h approaches zero, all the terms except $4a^3$ approach zero, so we conclude that $v = 4a^3$ at the instant $t = a$. Thus, for the law $s = t^4$, the velocity is given by the law $s' = 4t^3$.

Suppose we collect the results we have found.

$$\textit{Law for } s: \quad t^2 \quad t^3 \quad t^4 .$$

$$\textit{Law for } s': \quad 2t \quad 3t^2 \quad 4t^3 .$$

Most students notice the following. In the law for s' the exponent of t is one less than in the law for s. For example, when $s = t^{17}$, we expect s' to contain t^{16}. So, quite generally, for $s = t^n$, we expect s' to contain t^{n-1}. The other number in the formula is even easier to guess; we notice that $s = t^2$ leads to $s' = 2t$; $s = t^3$ leads to $s' = 3t^2$. The number written first in the formula for s' is just copied from the exponent in the formula for s. Thus for $s = t^{17}$, we expect $s' = 17t^{16}$. Quite generally, we guess that $s = t^n$ will give $s' = nt^{n-1}$.

This then is the answer to Exercise 4. It is an important result. So we set it out as formula.

Formula (4)   If $s = t^n$, then $s' = nt^{n-1}$.

Exercise 4 also suggested that we check our guess by testing it for $n = 1$. When $n = 1$, t^n is simply t, and we know that $s = t$ is the law for a body moving with velocity 1, so s' ought to be 1. Does the formula make it so? If it does not, our guess has been faulty. Putting $n = 1$ in formula (4), we get $s' = 1 \cdot t^0$. Here we have t^0. Some readers will know what this means, others will not, so I had better discuss it briefly. Consider the following expressions:

$$t^5, \quad t^4, \quad t^3, \quad t^2, \quad t, \quad 1, \quad \frac{1}{t}, \quad \frac{1}{t^2}.$$

Each expression is derived from the previous one by dividing by t. But if we look at the exponents, these begin 5, 4, 3, 2. The exponent falls by 1 at each step. This *suggests* that the expressions might also be written

$$t^5, \quad t^4, \quad t^3, \quad t^2, \quad t^1, \quad t^0, \quad t^{-1}, \quad t^{-2}.$$

So it seems that t^0 should mean 1. Then $s' = 1 \cdot t^0$ will mean $s' = 1 \cdot 1 = 1$, and this is what we hoped it would be.

The argument we have just had gives a meaning to t^0, t^{-1}, and t^{-2}. Does formula (4) work for these powers also? Can we, say, put $n = 0$ or $n = -1$ in formula (4) and get a correct result? This is a sheer guess, but let us try it out.

Putting $n = 0$ in formula (4) would give us the statement, "If $s = t^0$, $s' = 0 \cdot t^{-1}$." Is this a true statement? As we saw above, t^0 means 1. Also t^{-1} means $1/t$, but that does not matter much, because it gets multiplied by zero. So the result is zero. (I assume that t does not have the value zero; that would lead to all kinds of complications.) The statement thus becomes, "For the law $s = 1$, $s' = 0$."

What does this mean? $s' = 0$ means the velocity is zero; that is, the object is at rest. What does "the law $s = 1$" mean? It means that at *all times*, the object is distant 1 from some fixed place. If I keep my car all day just 1 foot from the door to my garage, surely it is at rest. The record of its progress is shown in Fig. 14. The graph neither rises nor falls. Its steepness or slope is zero: $s' = 0$.

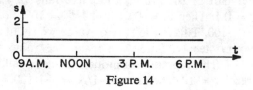

Figure 14

Now, what about $n = -1$? By comparing the two sequences,

$$t^5, \quad t^4, \quad t^3, \quad t^2, \quad t, \quad 1, \quad \frac{1}{t}, \quad \frac{1}{t^2},$$

$$t^5, \quad t^4, \quad t^3, \quad t^2, \quad t^1, \quad t^0, \quad t^{-1}, \quad t^{-2},$$

we were led to think that t^{-1} should mean $1/t$, and t^{-2} should mean $1/t^2$. If we put $n = -1$ in formula (4), we obtain the result that the motion $s = t^{-1}$ should have the velocity $s' = (-1) \cdot t^{-2}$. We can translate this by using the meanings for t^{-1} and t^{-2}. We thus reach the statement: to the law $s = 1/t$ there should correspond the velocity $s' = -1/t^2$. Is this reasonable? The graph of $s = 1/t$ is shown in Fig. 15. The graph falls very steeply at first, then moderately; it continues

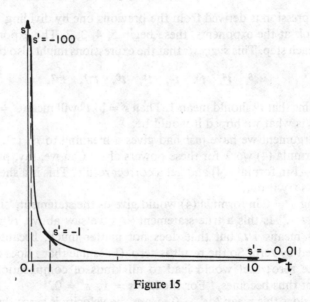

Figure 15

to fall, but it reaches a stage where it falls so gradually that you might almost think it was flat. Does our formula for s' agree with this type of behavior? The graph always falls. This means that s' should be negative throughout; and it is so. Whatever value of t you may choose, $s' = -1/t^2$ and s' is always negative.

In the earlier part of the graph, the curve falls very steeply. Suppose we try $t = 0.1$. Then $t^2 = 0.01$ and $1/t^2 = 100$. So, for $t = 0.1$, $s' = -1/t^2 = -100$. This is the sort of value we should expect for a graph falling very steeply. Later, we reach the value $t = 1$. Then $s' = -1/t^2 = -1$. This represents a moderate rate of fall, as we hoped. Still later we reach $t = 10$. Then $s' = -1/t^2 = -1/100 = -0.01$. This represents, again as we hoped, an almost imperceptible rate of fall.

Our guesses seem to be working out well. We guessed first that we could find a satisfactory meaning for t^n even when n was -1 or -2, and we guessed second that formula (4) could still be used with such values of n. On the basis of these guesses, we predicted the steepness at various places on the graph of $s = 1/t$, and we got results in good agreement with the actual appearance of this graph.

We often do this in mathematics—we take more than we are strictly entitled to. We may know that a method works in certain circumstances; we then experiment to see if, perhaps, it may not work in other circumstances as well; we try to push our laws out and out, until they cover as many cases as possible. Let us carry this idea a bit further.

Earlier we considered the sequence t^5, t^4, t^3, ..., in which we divide by t at each step. What happens if instead we write down a power of t and divide again and again by $\sqrt{t}$? We would obtain a series such as

$$t^2, \quad t\sqrt{t}, \quad t, \quad \sqrt{t}, \quad 1, \quad 1/\sqrt{t}, \quad 1/t, \quad \ldots.$$

Some of these we already know how to express in the form t^n. We write this information in, leaving gaps for results still unknown:

$$t^2 \quad t\sqrt{t} \quad t \quad \sqrt{t} \quad 1 \quad 1/\sqrt{t} \quad 1/t$$
$$t^2 \quad \cdots \quad t^1 \quad \cdots \quad t^0 \quad \cdots \quad t^{-1}.$$

In the lower row, we see the numbers 2, 1, 0, -1 with spaces between. What numbers should go in the spaces? The entries in the upper row are obtained by doing one operation again and again; divide by $\sqrt{t}$, divide by $\sqrt{t}$, divide by $\sqrt{t}$. This suggests that the numbers in the lower row may also be obtained by doing some operation again and again. The numbers 2, 1, 0, -1 go down by equal steps. Thus it is suggested that the operation is one of repeated subtraction. What must we subtract at each step, if we are to have 2, 1, 0 and -1 at the first, third, fifth, and seventh places? Clearly we must subtract 1/2, and the full sequence will be 2, 3/2, 1, 1/2, 0, $-1/2$, -1. This guess seems to give the simplest and most natural way of filling the gaps. We are thus led to the following table:

$$t^2 \quad t\sqrt{t} \quad t \quad \sqrt{t} \quad 1 \quad 1/\sqrt{t} \quad 1/t$$
$$t^2 \quad t^{3/2} \quad t^1 \quad t^{1/2} \quad t^0 \quad t^{-1/2} \quad t^{-1}.$$

We can now attach a meaning to t^n when n has one of the values 3/2, 1/2, $-1/2$.

What happens if we try putting $n = 1/2$ in formula (4)? Does it give sensible results? It would lead us to believe that, when $s = t^{1/2}$, the velocity $s' = (1/2)t^{-1/2}$; that is to say, when $s = \sqrt{t}$, we obtain $s' = (1/2)(1/\sqrt{t}) = 1/2\sqrt{t}$. Is this reasonable? The graph of $s = \sqrt{t}$ for positive values of t is shown in Fig. 16.

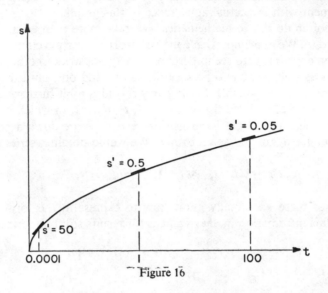

Figure 16

What do we expect for s' from looking at the graph? The graph rises throughout; we expect s' to be positive for all positive values of t. This is so, if $s' = 1/2\sqrt{t}$. The graph rises steeply at first, then moderately, and finally the slope becomes very gentle indeed. So we expect s' to be large at first, then of medium size, and finally very small. The formula $s' = 1/2\sqrt{t}$ agrees with these expectations. If you take for example $t = 0.0001$, then $\sqrt{t} = 0.01$ and $s' = 50$. For $t = 1$, $s' = 1/2$. For $t = 100$, $\sqrt{t} = 10$ and $s' = 1/20 = 0.05$.

It thus seems that we get correct results by using formula (4) even when n is fractional or negative. This is in fact true and can be proved. I shall not try to prove it here. A strict, logical proof would have to approach things in quite a different order. We should not grope our way forward from x^2 and x^3 to x^{-1} and $x^{1/2}$ by guesswork; we should begin with a definition of x^n that held equally well whether n was 2 or -1 or $1/2$ or $\sqrt{2}$ or π. From this definition we should deduce formula (4), so that all possible cases would be dealt with at one blow. But this proof would use some calculus ideas; you would certainly not see what the proof was aiming at, unless you had some familiarity with calculus already. Since we are not going to give a genuine proof, it seems better not to give a phony argument that appears to be a proof and is not. It is better to say plainly—we have made some guesses; we have produced evidence to show that these guesses lead to some reasonable results. They are in fact correct.†

In high-school algebra, students usually meet $t^{1/2}$ and t^{-1} before they know anything about calculus. Fractional and negative exponents may seem rather futile and aimless. The work we have just done shows the value of such exponents. The three expressions t^2, $1/t$, $\sqrt{t}$ look entirely different. If you were asked to find the velocity corresponding to each of them, you would think you had three quite different problems. But if you know enough about exponents to see these three expressions as t^2, t^{-1}, $t^{1/2}$, the three apparently distinct problems merge into one— they are all special cases of finding s' for $s = t^n$. Formula (4) covers them all: we solve all three problems at one blow. Thus our work on fractional and negative indices bears useful fruit; it saves us the trouble of learning three separate rules for these three cases. There are many places in calculus where it will have a similar labor-saving effect.

† For the benefit of students who intend to follow the subject further, I ought to indicate the line of proof I have in mind. It is possible to define log x by means of the integral calculus, and to deduce the properties of logarithms and antilogarithms. x^n can then be defined as antilog (n log x).

CHAPTER FIVE

Extending Our Results

Let us look back for a moment and take stock. What have we learned in all these pages? Not much! A good deal of our discussion has been concerned with rather general ideas—that calculus studies speed, velocity, rates of growth; that there are difficulties in thinking about variable velocity; that graphs can help us to visualize our problems. When you ask what exactly we have learned to calculate, the answer can be given very briefly—we have arrived at formula (4). We have learned how to calculate the velocity s' for the law $s = t^n$. Just that.

Now of course it is rare that we have to deal with such a simple formula as $s = t^n$. Mathematical and scientific formulas alike tend to be rather more complicated than this. However, $s = t^n$ gives us a kind of building block, from which many more elaborate formulas can be constructed. For example, $s = 16t^2$ is the law for a stone falling from rest under gravity. If a stone is thrown upwards with a beginning speed of 40 feet a second, its height after t seconds is given by the formula† $s = 40t - 16t^2$. In these formulas, powers of t occur, but other things occur as well; we cannot find s' by using formula (4) alone; we must know how to deal correctly with the minus sign and the numbers 40 and 16 in $s = 40t - 16t^2$, and with the number 16 in the formula $s = 16t^2$.

It is therefore natural to seek for principles that will allow us to

† The formulas $s = 16t^2$ and $s = 40t - 16t^2$ could be obtained experimentally by observing actual stones. In practice, it is more convenient to make certain other experiments, guess certain general laws of mechanics, and deduce these particular results mathematically.

answer questions such as "What is s' for $s = 5t^3 - 11t^2 + 8t + 9$?"
Here we have a simple expression, such as we meet frequently in school
algebra. It is quite easy to find s' for any such expression.

In discussing this question, it is convenient to use an idea mentioned
on page 18; we think of s' as being the *rate of growth of s*. Now imagine
a bar made up of two parts, of lengths y and z inches (Fig. 17).

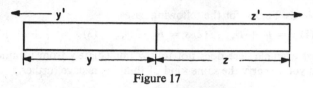

Figure 17

These parts are perhaps made of different metals; the metals are being
heated and so each part is expanding. What would be appropriate
symbols for the rates at which they are expanding? The first part is
of length y inches. y' is the natural symbol for representing the rate at
which y is growing. In the same way, z' represents the rate at which z
is growing. Suppose we know these numbers y' and z'. How would
you find out the rate at which the total length of the bar is growing?
Most people say at once, "I would add the rates at which the two parts
are growing." That is, $y' + z'$ gives the rate at which the total length is
growing. *The rate at which the whole grows can be found by adding
together the rates at which the parts grow.*

One can think of many illustrations. For example, if we use m for
the total number of males in the world and f for the total number of
females, we shall naturally use m' for the rate at which the male population is increasing, and f' for the rate at which the female population is
increasing. If w stands for the total population of the world, then
$w = m + f$, and w is increasing at the rate $w' = m' + f'$. You can devise
other examples for yourself.

For purposes of future reference we will record this simple principle
in the shape of a formula:

Formula (5)
$$\text{For } s = y + z,$$
$$s' = y' + z'.$$

But, for goodness' sake, do not learn this formula by rote. Think
through for yourself several times the way in which we arrived at this
result. You should then remember it without any effort.

We can apply our principle to make calculations. For example, if
$s = t^2 + t^3$, what is s'? Here s is the sum of two parts, t^2 and t^3. We
found earlier the rates of growth of these two parts. We know that t^2

grows at the rate $2t$, and t^3 grows at the rate $3t^2$. The rate of growth of the whole is found by adding these together. So we have:

$$\text{For } s = t^2 + t^3,$$
$$s' = 2t + 3t^2.$$

EXAMPLES. Find s' for the following cases.

(1) $s = t^2 + t^4$,　　(2) $s = t^3 + t^6$,　　(3) $s = t^2 + t^5 + t^7$.

(The last example here does not follow immediately from formula (5), but requires you to carry the same kind of thinking a stage further.)

A very similar principle holds for expressions involving minus signs. Suppose a man has y dollars to his credit in the bank, but debts amounting to z dollars. His wealth is then represented by $y - z$ dollars. If his bank credit is increasing at the rate of y' dollars a day and his debts are increasing at z' dollars a day, at what rate is his wealth increasing? Most people arrive, without difficulty, at the answer $y' - z'$, and are prepared to admit the reasonableness of the formula below.

Formula (5a)
$$\text{For } s = y - z,$$
$$s' = y' - z'.$$

By an extension of such ideas, we can deal with formulas containing several plus and minus signs. For example, we have:

$$\text{For } s = t^2 - t^3 + t^4 + t^5 - t^6,$$
$$s' = 2t - 3t^2 + 4t^3 + 5t^4 - 6t^5.$$

We now know how to deal with $t^2 + t^3 - t^7$, but we cannot yet find s' for $s = 5t^2$ or any similar expression. We need a further principle to deal with such cases.

Imagine a plant growing, and a lamp so placed that the shadow of the plant falls on the wall. It is possible to arrange the lamp in such a way that the shadow of the plant is always five times as high as the plant itself (Fig. 18). If h is the height of the plant in inches, the shadow will be $5h$ inches high. The plant is growing. h' represents the rate at which h is growing. How fast is the shadow growing? The shadow is always 5 times as high as the plant. If the plant grows by one inch, the shadow must grow by 5 inches. It seems reasonable that the shadow must be growing just 5 times as fast as the plant. Its rate of growth must then be $5h'$. So we arrive at the following conclusion:

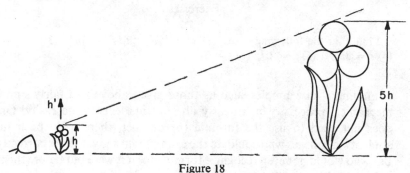

Figure 18

When h is growing at the rate h',
$5h$ is growing at the rate $5h'$.

Now there is nothing special about the number 5. If the lamp had been so placed that the shadow was 3 times as high as the plant, we should have arrived at the conclusion

$3h$ grows at the rate $3h'$,

and by shifting the lamp's position we could arrive at a whole series of statements such as

$2h$ grows at the rate $2h'$,
$4h$ grows at the rate $4h'$,
$7h$ grows at the rate $7h'$.

Here we have a whole series of arithmetical statements, obviously crying out to be replaced by a single algebraic statement. If instead of choosing particular numbers 5, 3, 2, 4, 7, ... more or less at random, we use the single algebraic symbol c, we shall be able to write all the statements given above simultaneously in the form:

Formula (6)   When h is growing at the rate h',
$c \cdot h$ is growing at the rate $c \cdot h'$.

It is understood here that c stands for any fixed number you may like to choose, such as 5, 3, 2, 4, 7 etc.

We are now able to deal with the question, "What is s' for $s = 5t^2$?" For $5t^2$ is always just 5 times as big as t^2. We know that t^2 grows at the rate $2t$; so $5t^2$ grows at the rate $5 \cdot 2t$, that is to say, at the rate $10t$. Hence, if $s = 5t^2$, $s' = 10t$.

Another example: what is s' for $s = 7t^4$? t^4 grows at the rate $4t^3$. So $7t^4$ grows at the rate $7 \cdot 4t^3$, that is, $28t^3$. So, if $s = 7t^4$, $s' = 28t^3$.

Exercises

Find s' in the following cases. (1) $s = 10t^2$, (2) $s = 20t^3$, (3) $s = 4t^4$, (4) $s = 8t^{100}$, (5) $s = 2t^3$, (6) $s = 3t^2$.

Finding s' in examples such as those given above is a fairly simple mechanical process, which nearly all students learn quite easily. One does not need to use the formula (6). In fact, there must be many students in colleges who could do the examples just given quite correctly, but who would not even recognize formula (6) if it were put before them.

I could show you how to find s' for $s = 5t^2$ without ever mentioning formula (6). In fact, I have done so; if you will look at the paragraph beginning "We are now able to deal with the question: What is s' for $s = 5t^2$?" you will see that it makes no mention of the formula (6). Why then did I bother to mention formula (6) at all? Mainly for purposes of easy reference; if at any future time I shall say, "By formula (6)," I shall *not* mean that I want you to write down formula (6) and make some mechanical substitution into it. Rather this will be shorthand for a sentence like the following: "Do you remember the picture of the plant and its shadow, and how we used that picture to think out the rate of growth of $5t^2$? Well, think along those lines now." In the same way, "By formulas (5) and (5a)" will be used to remind you of the way we thought about the metal bars joined end to end, and to suggest that the same line of thought will enable you to understand whatever problem or process it is that I am discussing. Indeed throughout mathematics, this is what formulas should mean to you—not a recipe to be carried out blindly, but to remind you that here is another example of something that you have already studied and understood.

In practice, the formulas (or ideas) (5) and (6) are usually combined. For example, we may want to find s' for

$$s = 5t^2 + 7t^4.$$

Here s breaks up into two parts, $5t^2$ and $7t^4$. Formula (6) tells us how quickly each of these parts grows; in fact we calculated these rates of growth on page 43. Formula (5) tells us that we can find the rate of growth of s by adding together the rates of growth of the two parts.

The argument might be written out as follows—

By (6), $5t^2$ grows at the rate $10t$.

By (6), $7t^4$ grows at the rate $28t^3$.

We can combine these results by using (5). Therefore $5t^2 + 7t^4$ grows at the rate $10t + 28t^3$.

In practice, finding such rates of growth is a very simple mechanical process, and you never see the argument written out in full.

Exercises

Find s' in the following cases: (1) $s = 10t^2 + 20t^3$, (2) $s = 2t^3 + 3t^2$, (3) $s = 5t^7 + 2t^4$, (4) $s = 5t^7 - 2t^4$, (5) $s = 10t^2 + 20t^3 - 5t^4$.

Rate of Growth of Any Polynomial

An expression such as $10t^4 + 7t^3 - 3t^2 + 5t + 11$ is called a *polynomial*. All the expressions used in recent examples have been polynomials. In fact, the information contained in formulas (4), (5), (5a), (6) is sufficient to enable us to find the rate of growth of any polynomial.

It is a curious fact that the simplest part of the work causes the largest number of errors. Most students can deal with the higher powers. They begin quite happily—

$$10t^4 \text{ grows at the rate } 40t^3,$$
$$7t^3 \text{ grows at the rate } 21t^2,$$
$$3t^2 \text{ grows at the rate } 6t.$$

Then the troubles seem to begin. Many students are puzzled to know how fast $5t$ grows. And the constant term, 11, gives most trouble of all.

If you will look back to Chapter 2 in the section "Finding velocity in simple cases," you will find that we worked out the rate of growth of various simple expressions. Result A, for instance, shows that $10t$ grows at the rate 10. On page 20 I hope you discovered for yourself that $20t$ grows at the rate 20; that $30t$ grows at the rate 30, and so on. In Exercise 5 of page 20, I hope you managed to collect the results of earlier questions into the algebraic generalization that kt grows at the rate k. So, in our question, $5t$ grows at the rate 5.

The constant term 11 is the simplest of all. 11 is a fixed number. It does not grow at all. If $s = 11$, $s' = 0$. If you look back again to pages 19 through 21, you will see that the constant term contributed *nothing at all* to the final answer. Compare for example

$$Result\ A\text{: }when\ s = 10t,\ s' = 10$$

with

$$Result\ B\text{: }when\ s = 10t + 3,\ s' = 10.$$

The $+3$ in $s = 10t + 3$ makes no difference at all in the answer for s'. If one car were moving according to the law $s = 10t$ and another car were moving according to the law $s = 10t + 3$, the second car would always be exactly 3 feet ahead of the first car. The distance between them does not change. This means that they are both moving *at the same speed*. That is why, although the laws for s are different, the laws for s', in Results A and B, are the same.

Accordingly, the full argument for the problem, to find s' for $s = 10t^4 + 7t^3 - 3t^2 + 5t + 11$, is the following:

$$10t^4 \text{ grows at the rate } 40t^3,$$
$$7t^3 \text{ grows at the rate } 21t^2,$$
$$3t^2 \text{ grows at the rate } 6t,$$
$$5t \text{ grows at the rate } 5,$$
$$11 \text{ grows at the rate } 0.$$

These are the rates at which the various parts grow; combining these in accordance with the principles of formulas (5) and (5a), we find that $10t^4 + 7t^3 - 3t^2 + 5t + 11$ grows at the rate $40t^3 + 21t^2 - 6t + 5 + 0$. Thus $s' = 40t^3 + 21t^2 - 6t + 5$.

In practice, it is rather more convenient to write the rates of growth of the various parts underneath the formula for s, like this:

$$s = 10t^4 + 7t^3 - 3t^2 + 5t + 11$$
$$40t^3 \quad 21t^2 \quad 6t \quad 5 \quad 0.$$

We can then form the expression for s' simply by copying down the plus and minus signs from the formula for s, thus

$$s = 10t^4 + 7t^3 - 3t^2 + 5t + 11$$
$$s' = 40t^3 + 21t^2 - 6t + 5 + 0.$$

Of course, the final $+0$ makes no difference to the answer. Beginners, however, would be wise to put it in, until they are expert at this type of operation. The commonest mistake that students make is simply to copy down the 11, so that they obtain the *incorrect answer*,

$$s' = 40t^3 + 21t^2 - 6t + 5 + 11.$$

This is quite wrong. 11 does not grow at the rate 11. The number 11 stays the same always; it does not change; it does not grow; its rate of growth, therefore, is zero.

You are less likely to make this mistake if you think what you are doing. Underneath each term in the expression for s you are writing *the rate at which that part grows;* you are then combining these to find the rate at which the whole grows. A constant, such as the term 11 in our example, represents a part of fixed size, a part whose rate of growth is zero:

$$\begin{array}{llllll} \text{Time, } t & 0 & 1 & 2 & 3 & 4 \\ \text{Size of part} & 11 & 11 & 11 & 11 & 11. \end{array}$$

On the other hand, the term $5t$ represents a part that is growing:

$$\begin{array}{llllll} \text{Time, } t & 0 & 1 & 2 & 3 & 4 \\ \text{Size of } 5t & 0 & 5 & 10 & 15 & 20. \end{array}$$

The part represented by $5t$ is growing steadily at the rate 5.

It is wise at first to work slowly. Do not be afraid to spend quite a lot of time thinking about the difference between a variable term like $5t$ and a constant term like 11. Make tables like those above; or draw pictures showing a bar of fixed length, 11 inches, joined to an expanding bar, of length $5t$ inches. However slowly you go, make sure that you are talking sense. As you continue with the work, you will form correct habits and, without your noticing it, your speed of work will increase. There are plenty of students who can write down the *wrong* answer quickly! Their work has no value at all.

Exercises

1. If $s = 12$, what is s'? If $s = 2t$, what is s'? If $s = 2t + 12$, what is s'?
2. If $s = 7$, what is s'? If $s = t^3$, what is s'? If $s = t^3 + 7$, what is s'?
3. If $s = 8$, what is s'? If $s = 3t$, what is s'? If $s = t^2$, what is s'?
   If $s = t^2 + 3t + 8$, what is s'? Find s' in the following cases:
4. $s = 5t^2 + 4t + 3.$
5. $s = 5t^2 - 4t + 3.$
6. $s = 2t^3 - 3t^2 - 10t + 100.$
7. $s = 4t^{20} + 2t^{15} - 3t^{10} + 5t + 17.$
8. $s = 10t^6 + 12t^5 - 15t^4 + 20t^3 - 30t^2 + 60t + 60.$

You may need to practice with further examples of this type, to acquire speed and accuracy of working. But so far as ideas are concerned, there is nothing more to say; if you understand how to do all the exercises above, you have mastered this particular topic. If you are given any polynomial for s, you can find its rate of growth s'.

An Application of s'

On page 40 it was mentioned that, if a stone is thrown vertically upwards with a beginning speed of 40 feet a second, its height thereafter (so long as it remains in the air) is given by the law $s = 40t - 16t^2$. We know from experience that the stone would rise at first, after a certain time would come to rest, and then would begin to fall. We might ask questions such as the following. (1) How long does the stone continue to rise? (2) At what time does it come to rest at the top of its path, just before it begins to fall? (3) What is its velocity when it has been in the air for 1 second? (4) What is its velocity when it has been in the air for 2 seconds?

It would be possible to answer some of these questions in a primitive way without using calculus. By making a table and perhaps sketching a graph one could find—after a certain amount of guesswork—how long the stone continued to rise, and when it reached its highest point. For questions (3) and (4), which are concerned with velocity, we should have to go through the rather tedious arithmetical process that we used earlier for estimating velocities.

All four questions have something to do with velocity, so calculus is the natural method to use.† The calculus approach is simple and uses very little in the way of calculation.

First of all, we obtain a formula for the velocity s'. Since $s = 40t - 16t^2$, it follows that $s' = 40 - 32t$. We answer question (2) first; when does the stone come to rest before it starts on its downward path? When the stone is at rest, its velocity is zero; that is, $s' = 0$. When is s' zero? Since $s' = 40 - 32t$, we have to find when $40 - 32t$ equals zero. So we have the equation $40 - 32t = 0$. This is easily solved and gives $t = 1\frac{1}{4}$. So the stone reaches its highest point after $1\frac{1}{4}$ seconds.

Question (3); what is the velocity after 1 second? In algebraic terms, what is s' when $t = 1$? This is just a matter of substituting $t = 1$ in the equation $s' = 40 - 32t$. We find that $s' = 8$ when $t = 1$. This t gives the velocity of the stone after 1 second, namely 8 feet a second.

Question (4); what is the velocity after 2 seconds? We substitute $t = 2$ in $s' = 40 - 32t$ and find $s' = -24$. There is a minus sign in this last answer; what does that mean? We discussed the interpretation of negative velocity on page 17. Positive velocity indicates that the stone is rising; negative velocity that it is falling. We have had one of each; for $t = 1$, we found $s' = 8$, indicating that the stone was rising at a speed of 8 feet a second; for $t = 2$, we found $s' = -24$, indicating that the stone is now falling at a speed of 24 feet a second.

† Some students avoid calculus by quoting formulas from mechanics, but as calculus methods give the simplest way of proving these formulas in mechanics, this does not really make much difference.

These results are reasonable. If the stone reaches its highest point at $t = 1\frac{1}{4}$, we should expect it to be rising at all times before $t = 1\frac{1}{4}$ and falling at all times after $t = 1\frac{1}{4}$. Because $t = 1$ comes before $t = 1\frac{1}{4}$ and $t = 2$ comes after $t = 1\frac{1}{4}$, all our results fit together to make a reasonable and consistent picture.

In discussing the other questions we have just about answered question (1). The stone rises between $t = 0$ and $t = 1\frac{1}{4}$. This checks with our algebraic information. The equation $s' = 40 - 32t$ could be written as $s' = 32 \cdot (1\frac{1}{4} - t)$. So long as t is less than $1\frac{1}{4}$, the value of s' will be positive, and so the stone will be rising.

Here we have used a simple formula to answer a simple question. In the application of calculus to mechanics and astronomy, much more complicated formulas and much harder problems arise. But this example may give a faint indication of the way in which calculus is used in scientific applications.

CHAPTER SIX

Calculus and Graphs

Earlier in this book, we saw that there was a close connection between movement and curves. A moving object could be made to leave an inked trail on the paper; this curve then gave us a record of how the object had moved. By passing the curve past a narrow slit, we could again see the movement of the point rising and falling. Alternatively, we could make a cam to the shape of the curve. (See pages 13 and 14.)

The curve is a complete record of the motion. Anything that can be said about the motion can be deduced by examining the curve.

Until now, we have spoken mainly in terms of motion. We have thought of s' as measuring the velocity of a moving object. But the velocity of the object at any moment must somehow or other be shown by the shape of the corresponding curve. So it ought to be possible to interpret s' as describing some geometrical property of the curve.

We have already touched on this question twice (pages 13 and 22). We came to the conclusion that the velocity of the object was related to the steepness of the curve. So s' should measure the steepness of a curve. The general idea here is clear enough. If s' is large, say $s' = 100$, we should be dealing with a very steep curve; if s' is small, say $s' = \frac{1}{4}$, the curve should be not very steep. If $s' = 0$, the curve should be flat. But "very steep" and "not very steep" are rather vague terms. On the other hand, the values of s' are perfectly precise. There is no vagueness at all in saying $s' = 100$ or $s' = \frac{1}{4}$. Is there any way in which we can measure the steepness of a graph with the same perfect precision? To answer this question, we look through our earlier work and try at each step to interpret in terms of graphs rather than in terms of moving objects.

We began our investigation of velocity by thinking about constant velocity. When a body moves with constant velocity, the corresponding graph is a straight line. We now follow through the argument used to derive formula (1), on page 15.

We began there with the little table

$$
\begin{array}{ccc}
t & a & b \\
s & p & q
\end{array}
$$

The information contained in this table appears on the graph in Fig. 19.

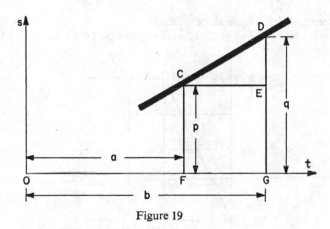

Figure 19

The line *CD* records the motion of the object. The point *C* on the graph has co-ordinates (a, p) and shows that when $t = a$, $s = p$. In the same way, the fact that the graph passes through the point *D* shows that when $t = b$, $s = q$. In formula (1) we used v to measure the velocity. Since then, we have become used to the symbol s' so we may rewrite formula (1) as

$$
s' = \frac{q - p}{b - a}.
$$

This result was originally obtained by considering velocity as "distance divided by time". Can we interpret the above equation geometrically? Can we find lines whose lengths are $q - p$ and $b - a$ and consider their ratio?

This is not a difficult problem. Since $OF = a$ and $OG = b$, evidently $FG = b - a$. As *FGEC* is a rectangle, the lengths *CE* and *FG* are equal. So, as a geometrical interpretation of the number $b - a$, we can use either the length *FG* or *CE*, whichever we find more convenient. Again, the lines CF and EG are both of length p. As $DG = q$, we have $q - p = DG - EG = DE$. Hence

$$\frac{q - p}{b - a} = \frac{DE}{CE}.$$

The ratio of DE to CE does in fact give us a way of measuring the steepness of the line CD. The larger this ratio is, the steeper the line will be. Accordingly, we will adopt this ratio as a measurement of steepness, and refer to it as the *slope* of the line.

EXAMPLE. What is the slope of the line $y = 2x + 1$? The graph of $y = 2x + 1$ is shown in Fig. 20. Choose any two points on the line for C and D.

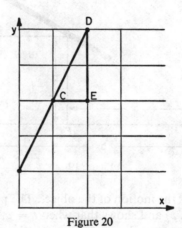

Figure 20

I have chosen $(1, 3)$ and $(2, 5)$. Then $CE = 1$ and $DE = 2$. Therefore $DE/CE = 2/1 = 2$. Any other pair of points will lead to the same result. The slope of the line is 2.

Exercises

Find the slopes of the lines (1) $y = x$, (2) $y = x + 1$, (3) $y = 2x$, and (4) $y = 3x$.

In all the examples given above, the slope turns out to be a positive number. But of course the graph may be like Fig. 21. In this case, $a = 3$, $b = 5$, $p = 4$, $q = 2$, and

$$\frac{q - p}{b - a} = \frac{2 - 4}{5 - 3} = \frac{-2}{2} = -1.$$

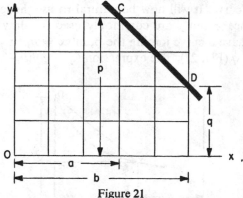

Figure 21

So this line has slope -1. We shall say that the line CD in Fig. 19 goes "uphill," the line CD in Fig. 21 goes "downhill." Whenever you have a line going downhill, you must expect a negative value for the slope, just as we got a negative value for the velocity of a falling object.

Exercises

Find the slopes of the following lines: (5) $y = 5 - x$, (6) $y = 10 - 2x$.

Slopes of Curves

When we were studying velocity, we began with the simple procedure that is used in arithmetic lessons; see how many miles the object has traveled; see how many hours it has taken to do this; divide the first number by the second. This procedure is briefly indicated by saying "velocity is distance divided by time." It leads to formula (1),

$$s' = \frac{q - p}{b - a}.$$

We then saw that this formula was not helpful for an object that moved in an irregular way, now stopping, now starting, now moving fast, now moving slow. For such an object, the total number of miles gone, divided by the total number of hours taken, gave us only the average velocity, which might be very different from the actual velocity at any instant of the journey. We got round this difficulty by considering average velocity over shorter and shorter intervals. If, as the interval became shorter and shorter, the average velocity approached some fixed value, we called that value the true velocity at an instant.

We can follow the same procedure when we try to find the slope at a point of a curve. It will now be natural to use the letters x, y rather than s, t, since x and y are commonly used for drawing graphs. We suppose we have a curve joining the point $x = a$, $y = p$ to the point $x = b$, $y = q$ (Fig. 22). The expression $(q - p)/(b - a)$ measures the

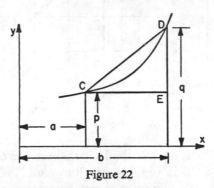

Figure 22

slope of the line CD. We can put this expression into words. Since D is at height q while C is at height p, the quantity $q - p$ measures the height risen in going from C to D. The quantity $b - a$ measures the length CE. If we move from C to D, the length CE tells us how far we have moved across the paper. If C and D were two actual places, $b - a$ would be the distance between them as shown on a map, because a map is made as if someone were looking down on a country from above. The height of a place does not affect where it appears on a map. Thus we could say that $(q - p)/(b - a)$ represents "height risen divided by map distance". For instance, if a person traveled 2000 miles East and rose through a height of 3 miles, the fraction $(q - p)/(b - a)$ would be 3/2000. This fraction involves the total height risen and the total distance traveled; it would not tell us anything about how steeply the traveler's plane was climbing at any moment. In the same way, in Fig. 22, the fraction $(q - p)/(b - a)$ gives the slope of the line CD; this does not coincide with the steepness of the curve either at C or at D.

However, for the curve shown in Fig. 22, it appears reasonable that if D were to approach closer and closer to C, then the slope of *the line*

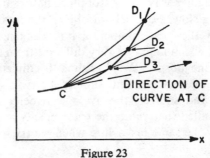

Figure 23

CD might approach closer and closer to the slope of *the curve at C* (see Fig. 23). D_1 is the first position we try for *D*; D_2 is the second position and D_3 is the third. The picture suggests that, the closer we take *D* to *C*, the closer will the line *CD* come to the true direction of the curve at *C*. But in this picture it is not possible to choose *D* so that the line *CD* actually coincides with the dotted line. We can get as near as we like, but we never actually arrive.

For example, suppose we want to find the slope of the parabola $y = x^2$ at the point where $x = 3$, as shown in Fig. 24. *C* is the point

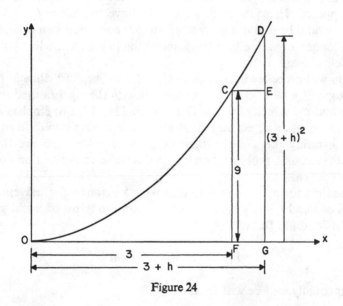

Figure 24

(3, 9), so $OF = 3$ and $FC = 9$. *D* is to be chosen somewhere near *C*. So we take $x = 3 + h$ for *D*. For the moment, we shall leave it undecided what *h* is going to be. We shall certainly choose fairly small values for *h*, because we want *D* to be close to *C*. So $OG = 3 + h$. Now what about *GD*? How do we graph the curve $y = x^2$? We choose any value for *x*; we square it; that gives us the value of *y*, which we measure vertically upwards. We have already followed this procedure for *C*. We measure the distance *FC* equal to 9, because *OF* equals 3, and 9 is the square of 3. In the same way *OG* equals $3 + h$, so the vertical distance *GD* must be made equal to the square of $3 + h$. Thus $GD = (3 + h)^2$.

Finding *DE*/*CE* is now straightforward.

$$CE = FG = OG - OF = (3 + h) - 3 = h,$$

$$DE = GD - GE = GD - FC = (3 + h)^2 - 9 = 6h + h^2.$$

Hence

$$\frac{DE}{CE} = \frac{6h + h^2}{h} = 6 + h.$$

So the slope of the line CD is $6 + h$. I can make this as near to 6 as I like by choosing h small enough. For example, if I want the slope to be 6.001, I can choose $h = 0.001$. The smaller h is, the closer will the slope of CD come to 6. So the number 6 comes quite clearly out of our work; it is indicated as being the slope of the dotted line. But we can never make CD actually coincide with the dotted line. For to make the slope of CD equal to 6, we should have to choose $h = 0$. Then $6 + h$ would be 6. But $h = 0$ means that D coincides with C, and there is no longer any sense in talking about the line CD; the line joining C to itself is meaningless.

You will probably have noticed that the algebra just done in finding the slope of $y = x^2$ at $x = 3$ is identical with the algebra used on page 29 to find the velocity for $s = t^2$ at $t = 3$. This helps to emphasize that the motion of an object and the shape of a curve are two different ways of illustrating the same mathematical idea. When you are thinking about a calculus problem, you can use whichever illustration you find more convenient.

Needless to say, all the formulas we have found for movement, in terms of s and t, apply equally well to graphs in terms of x and y. Thus our basic result, formula (4),

$$\text{If } s = t^n, \text{ then } s' = nt^{n-1},$$

could equally well be written as

$$\text{If } y = x^n, \text{ then } y' = nx^{n-1}.$$

All the examples we have done to calculate s' immediately give us corresponding results about y'.

The Extra Information Given by Calculus

If we plot the graph $y = x^2$ by the usual, elementary method, we just find points that the graph goes through. Putting $x = 3$ gives $y = 9$, so the graph goes through the point $(3, 9)$. But nothing is known about the direction in which it passes through this point; you just have to guess that by looking at the other points and seeing how the curve seems to run.

Calculus provides us with information about direction. If $y = x^2$, we know that $y' = 2x$. For $x = 3$, $y = 9$ and $y' = 6$. So the curve passes through the point $(3, 9)$ with slope 6.

A line of slope 6 is one that rises six units for every unit across. It appears as in Fig. 25. For our purpose we shall not need such a long

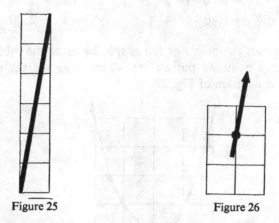

Figure 25   Figure 26

piece of line; a small sample of it will be quite sufficient to show the direction. Then instead of simply plotting the point $(3, 9)$ on the graph paper, we shall be able to plot a point and a small arrow, as shown in Fig. 26. The curve goes through the point $(3, 9)$ in the direction indicated by the arrow.

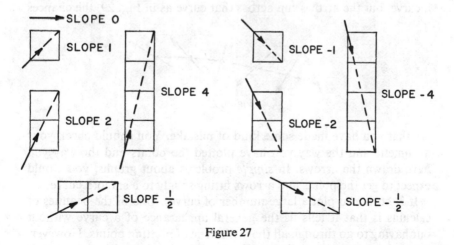

Figure 27

For drawing graphs we shall need arrows representing other slopes. A selection of these is shown in Fig. 27.

Now suppose we wish to plot the graph $y = x^2$ from $x = -2$ to $x = +2$. We first calculate the tables below:

$y = x^2$							$y' = 2x$				
x	-2	-1	0	1	2	x	-2	-1	0	1	2
y	$+4$	$+1$	0	1	4	y'	-4	-2	0	2	4

We then plot the points of the graph, by using the table for y, and through each point we put an arrow, by using the table for y'. We arrive at the diagram of Fig. 28.

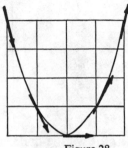

Figure 28

We then join the points by a curve; the direction of this curve, as it passes through each point, should agree with the direction of the arrow.

If, in any exercise of this kind, you obtain points that seem to lie on a curve, but the arrows run across that curve as in Fig. 29, the chances

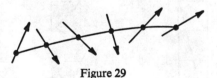

Figure 29

are that you have made some kind of mistake. You should check your arithmetic and the way you have plotted the points and the way you have drawn the arrows. In *simple* problems about graphs, you would expect to get the points and arrows fitting snugly to a smooth curve.

It is boring to plot a large number of curves. One of the beauties of calculus is that it tells us the general appearance of a curve without our having to go through all the arithmetic of plotting points. However, it is good to plot just one or two curves by the method just described, in order to make yourself thoroughly familiar with the meaning of y' as a slope measurer.

Exercises

1. If $y = 10x - x^2$, then $y' = 10 - 2x$. Plot the graph of $y = 10x - x^2$ from $x = 0$ to $x = 10$ by the primitive method, that is, using the values of x and y only. Then *use the values of y'* to insert the arrows. Check that these arrows do touch the curve.

2. In the graph of $y = x^2$, prepared as in Fig. 28, insert the points and arrows corresponding to $x = -1/2$ and $x = +1/2$. Check that these fit the curve.

3. Sketch the graph $y = 4x - x^2$ from $x = 0$ through $x = 4$ by the method described in the text; that is to say, by drawing the points and arrows first, and then drawing a smooth curve through them.

A possible misunderstanding should be guarded against here. Sometimes students set out to graph $y = x^2$ by this method. They find $y' = 2x$. Then they see this last expression, $2x$, and they think, "The graph of $2x$ is a straight line." So they draw a straight line, and regard this as the answer to the problem.

So it should be emphasized that all the work we did under the heading "The extra information given by calculus" was aimed at plotting the graph of $y = x^2$. You know that the graph of $y = x^2$ is a parabola. So what we get at the end of our work must be this same parabola. The graph of $y = x^2$ is something fixed; it does not depend on which student is drawing it, or what that student knows. And yet some students seem prepared to believe that when you draw this graph in an algebra class the correct answer is a parabola, and when you draw it in a calculus class the correct answer is a line. Perhaps students simply feel that one teacher is pleased when you draw a parabola and another teacher is pleased when you draw a straight line, so the students try to please everybody. But the aim of mathematics is not to please people. The aim of mathematics is to find out the truth for yourself and to know the truth as it really is. If you have once become convinced that the graph of $y = x^2$ is a curve, then in no circumstances should you be willing to draw it as a straight line.

The equation $y' = 2x$ does precisely what the heading of the section promises; it gives some extra information about the parabola $y = x^2$. The equation $y' = 2x$ does not in any way contradict the equation $y = x^2$. The equation $y = x^2$, taken by itself, allows you to plot points. The equation $y' = 2x$ then allows you to put little arrows through these points, showing the direction in which the parabola passes through these points. The two equations give different kinds of information. The quantity y tells you how high a point is above the axis OX; the quantity y' tells you the direction of the curve.

The same distinction can be made in terms of moving objects. We have the two equations $s = t^2$ and $s' = 2t$. The quantity s tells you where the object is; the quantity s' tells you how fast it is moving.

The confusion seems to arise in students' minds because they use sentences beginning with "it". They will ask, "How can *it* be t^2 when *it* is $2t$?" By using the word "it" they manage to mix up two quite different things. The *distance gone* is given by t^2; the *speed* is given by $2t$. Later on we shall meet yet a third quantity, the *acceleration*, so that there will be three possible meanings for "it." It will then be even more important to say what you are talking about, and to avoid the vague use of "it."

Really I suppose the students who draw straight-line graphs are answering a different question. If I ask you to draw the graph of $s = t^2$, I am really asking you to draw a graph from which *the position of the object at any time* can be read off. This graph is a parabola. But I could instead have asked the question: draw a graph from which I can read off the *speed* at any time. This is quite a different question. The velocity is given by $2t$, and the graph of *velocity* against *time* is a straight line. So a straight-line graph is the correct answer to this second question, but it is the wrong answer to the first question.

Both inside mathematics and outside of it, people seem to make mistakes because their minds flit from one question to another; they start to answer question A, and part way through their thinking they begin answering question B. The answer, needless to say, is nonsense. We all do this to some extent. An important part of mental training is to learn to avoid this. Most people rush to answer a question or solve a problem. But really it pays, before even attempting an answer, to fix firmly in your mind what the question is; jot down a sentence or two, giving the essence of the question; or make a little sketch, showing what the question means; if you can see in part what the answer is going to be, make a note of that too. That will save you many errors. For instance, the present section began by discussing how calculus helps us to plot the graph of $y = x^2$; you know something about the graph of $y = x^2$ already; you know that it is a parabola, or at the very least you know it is not a straight line, but something shaped rather like the letter U. Very well then; the final answer must be something like the letter U, and if the linear expression $2x$ turns up part way through the work, and it should occur to you that the graph of $2x$ is a straight line, you will not be misled. You know you are looking for something U-shaped, not for something straight. This general feeling of what you are looking for is of great importance in mathematics; we all make errors of calculation and of thought, and it is only because we have this shadowy idea of what to expect that we are able to detect these errors. At a certain stage after an error, the results are usually so ridiculous that we know we must have made a slip somewhere.

Graphs Without Plotting

The plotting of points is a tedious business, and even the method of plotting points with arrows is not much better, once the first novelty has worn off. As was mentioned earlier, calculus enables us to get a general idea of the graph of an equation without plotting any points at all, and without using squared paper.

The method depends on the remark made earlier, that the slope y' is positive when a curve is rising, zero when the curve is flat for a moment, and negative when the curve is falling (Fig. 30).

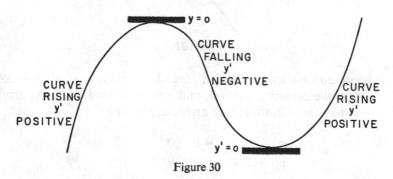

Figure 30

This can be illustrated from the graph of $y = x^2$ as shown in Fig. 28. $y' = 2x$, so y' is negative when x is negative; zero when $x = 0$; positive when x is positive. This agrees with the shape of the curve, which is falling so long as x is negative; is flat for a moment when $x = 0$; and rises when x is positive. Simply by examining the equation $y' = 2x$ it would have been possible to describe the general appearance of the graph. We would study the sign of y'. When is y' positive? Only when x is positive. So the slope is uphill only when x is positive. When is y' negative? Only when x is negative. So the slope is downhill only for negative x. When is y' zero? Only for $x = 0$. So the curve is flat only at $x = 0$.

This method is really more powerful than plotting points. We plotted the graph between $x = -2$ and $x = +2$. For all we know, all kinds of things may happen outside that interval. Between $x = 100$ and $x = 200$ the curve might twist and turn in a most complicated manner; plotting the values between -2 and $+2$ tells us nothing about what happens for large values of x. But the calculus method gives us this information. $y' = 2x$; and $2x$ is positive for every positive value of x. So we can be sure that the curve continues to climb however far we may go to the right. It always goes uphill, because y' is positive throughout this region. In the same way, we can be sure that the curve is downhill at all points to the left of the origin, because $y' = 2x$, and $2x$ is negative for all negative values of x.

We have used a certain convention in speaking of "uphill" and "downhill." We always think of ourselves as moving in the direction of increasing x, that is to say, from left to right. In our earlier diagram showing moving bodies this convention was used throughout. The earlier times were shown on the left, the later times were shown on the right. Thus a graph such as that of Fig. 31 represented a point moving *up* the page.

Figure 31

Suppose now we want to get an idea of the graph $y = 100x - x^2$. If we used the primitive method, and simply plotted the curve from $x = -2$ to $x = +2$, we should arrive at the table:

x	-2	-1	0	1	2
y	-204	-101	0	99	196

The numbers for y are steadily rising and if we plotted the curve simply on the basis of this evidence we might guess that the curve always went uphill. But in fact the significant things happen a long way from the interval studied above. Calculus tells us where things happen and what happens there. From $y = 100x - x^2$ we find $y' = 100 - 2x$. We may ask first, "Is the curve flat anywhere?" The curve is flat when y' is zero. So this leads us to examine whether we can choose x to make y' zero. We certainly can. $x = 50$ does the trick. That suggests that we examine what happens on either side of the flat position. If x is bigger than 50, $2x$ will be bigger than 100 and $100 - 2x$ will be negative. So the curve goes steadily downhill to the right of $x = 50$. In the same way, we can see that y' is positive whenever x is less than 50. So the curve is steadily uphill until $x = 50$ is reached. So we have an outline of the behavior of the curve, as shown in the following table:

Value of x	Less than 50	Equal to 50	More than 50
Value of y'	Positive	Zero	Negative
Meaning	Curve rising	Curve flat	Curve falling

This suggests that the curve may look something like that in Fig. 32.

This shows the general shape of the curve. But the curve is still very much up in the air. We have not shown the axes OX and OY at all. If

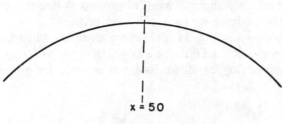

$$x = 50$$

Figure 32

we want to show how the curve lies in relation to the axes, we have to return, but only briefly, to the more primitive method. That is, we look at the original equation for y, not at the calculus result for y'. We plot one or two key points to tie the curve down. A certain judgment is called for, to decide which points will give useful information about the position of the curve, without involving too much arithmetic. Since the original equation $y = 100x - x^2$ may also be written in the factored form $y = x(100 - x)$, it is natural to consider the two values of x that make y zero, namely $x = 0$ and $x = 100$. It is also natural to consider $x = 50$, the value corresponding to the top of the hill. Taking the three values 0, 50, and 100 for x, we get the little table:

x	0	50	100
y	0	2500	0

This gives us three useful points on the curve, and in Fig. 33 we make a rough sketch of the graph with the help of this extra information.

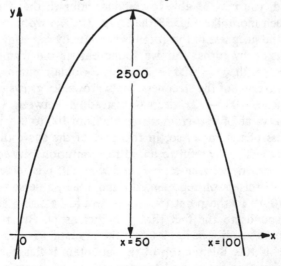

Figure 33

There is perhaps a slight element of guesswork in our drawing of the
curve. All the evidence we have collected is consistent with the shape
of the curve shown in Fig. 34. This curve also rises when x is less than
50, is flat for $x = 50$, and falls for x larger than 50. Also it goes through
the three points. So, for all we have yet proved, the graph might be

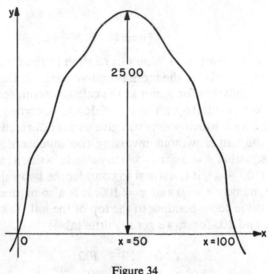

Figure 34

Fig. 34 and not Fig. 33. Later, we shall discuss a method that helps to
exclude the possibility of Fig. 34 being the graph. Without waiting for
this method, you may be able to convince yourself that the graph we
want is much more like Fig. 33 than Fig. 34. The wobbles in Fig. 34
mean that the steepness of the graph is perpetually wavering, increasing
and decreasing by turns. But we found earlier the formula for the
slope of the graph, $y' = 100 - 2x$. This formula contains nothing to
suggest a wavering of the steepness. As x grows, $2x$ grows steadily and
in consequence $100 - 2x$ decreases steadily. Between $x = 0$ and
$x = 50$, the value of y' *decreases steadily* from 100 to 0. y' measures
the steepness of the curve, so, in this part of the curve the steepness
gets steadily less. Be careful to avoid the confusion between y and y'
mentioned earlier. Between $x = 0$ and $x = 50$, y is increasing, y' is
decreasing. If this graph represented a mountain, a person going from
the point (0, 0) to the point (50, 2500) would be rising all the time.
This corresponds to the fact that y is increasing. But the climbing
would become steadily easier. At first the mountain is almost vertical,
the slope y' is 100. But the top of the mountain is flat, the slope y' is
zero. The fact that the slope becomes gentler as you proceed corresponds to the fact that y' is decreasing. In Fig. 34, as you go from

$x = 0$ to $x = 50$, you meet in turn easy and difficult pieces of climbing. In Fig. 34, the slope does *not* become steadily gentler in this section of the curve. So this graph does not correspond to the equation $y = 100x - x^2$. In the same way, if you investigate how the downhill steepness varies between $x = 50$ and $x = 100$, you will find that Fig. 33 gives a better picture of the situation than Fig. 34. The same type of argument, of course, could be used for the rest of the graph, which we have not shown in Fig. 33, that is, for negative values of x to the left and for values of x greater than 100 to the right.

As a rule, in sketching simple graphs, the best procedure is first to calculate y' and to see for what values of x the quantity y' becomes zero. One can then study what happens in between these values of x.

For example, we might wish to sketch $y = x^3 - 12x$. Here $y' = 3x^2 - 12$. When is y' zero? We form the equation $3x^2 - 12 = 0$ and solve it. We find the values $x = -2$ and $x = 2$. So we have the information:

$$x \quad \text{.......} \quad -2 \quad \text{..........} \quad 2 \quad \text{..........}$$
$$y' \quad \text{.......} \quad 0 \quad \text{..........} \quad 0 \quad \text{..........}$$

We now have three intervals to consider. What is y' like when x is less than -2? What is y' like when x is between -2 and 2? What is y' like when x is larger than 2?

It is natural to consider these intervals, because if y' is changing from positive to negative, you would expect it to pass through the value zero. This does not always happen; if you plot the graph of $y = 1/x^2$, you will find that the curve goes uphill (y' positive) for negative x and downhill (y' negative) for positive x (Fig. 35). So as x passes through the value zero, y' changes from positive to negative. But y' never takes the

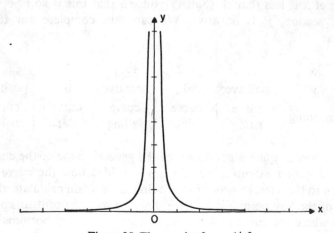

Figure 35. The graph of $y = 1/x^2$.

value zero; the curve is never flat. As we pass through $x = 0$, the curve jumps suddenly from being very steep uphill to being very steep downhill.

Such jumps then can occur with even such a simple expression as $1/x^2$. $y = \sqrt[3]{x^2}$ is another expression whose graph changes suddenly from steep downhill to steep uphill without having any flat part be-

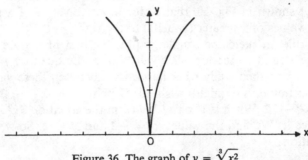

Figure 36. The graph of $y = \sqrt[3]{x^2}$

tween (Fig. 36). But the very simplest algebraic expressions, such as $y = x^2$ or $y = x^3 - 12x$, do not behave in this way. They do not jump, but creep from one situation to another. So, if y' is given by a simple formula of this kind (in technical language, if we are dealing with polynomials), the changes of value are gradual. If y' is going from positive to negative it must pass through the value zero; the same holds if it is changing from negative to positive.

Our question, with $y' = 3x^2 - 12$, illustrates this kind of behavior. We may write $y' = 3(x^2 - 4)$. If x is to the right of 2, or to the left of -2, x^2 will be bigger than 4, and y' will be positive. So y' is positive at the beginning and the end. But between $x = -2$ and $x = 2$, the square of x is less than 4. (Satisfy yourself that this is so.) So, in this middle section, y' is negative. We can thus complete our table as follows:

x	.....	-2	....	2	.....
y'	positive	0	negative	0	positive
meaning	curve rising	curve flat	curve falling	curve flat	curve rising

This gives us quite a good idea of the general shape of the curve. As with our earlier example, we still have no idea how the curve lies in relation to the axes. In order to tie it down, we again calculate and plot some of the key points. It will certainly be most helpful to know the places where the curve is flat—the hilltops and valley bottoms of this curve. So we calculate the value of y for $x = -2$ and for $x = 2$. It is easily seen from the equation $y = x^3 - 12x$ that the graph goes through the origin, for $x = 0$ gives $y = 0$. Are there any other points with $y = 0$? What are they?

EXAMPLE. Complete the investigation started above, and draw a sketch showing the curve $y = x^3 - 12x$.

In any such work, if you find yourself getting contradictory information, if the points and directions can only be fitted together by drawing an extremely complicated curve, it is wise to check and see if your work contains any errors. All the information from the various sources should fit neatly together to give a simple curve.

It may sometimes happen that, when we seek for places where $y' = 0$, we do not find any. Consider, for example, the graph of $y = x^3 + x$. Here $y' = 3x^2 + 1$. If we seek for places where y' is zero, by trying to solve the equation $3x^2 + 1 = 0$, we do not find any solutions.† Students are sometimes at a loss how to proceed. But the meaning is quite simple. There are no flat places on this curve; y' is never zero and never changes sign. Whatever value you care to choose for x, you get a positive value for y'. This means that the curve is always rising and in fact looks like the curve shown in Fig. 37. There is nothing

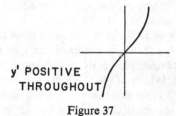

y' POSITIVE
THROUGHOUT

Figure 37

mysterious in the fact that we cannot find any solutions of the equation $y' = 0$. In fact, when you consider what the graph looks like, it would be very astonishing if we could find values of x to satisfy this equation. For such values would correspond to flat places on the curve, and there are none.

In all the examples considered so far, flat places, places where $y' = 0$, have occurred only at the tops of hills and at the bottoms of valleys. There is however another possibility. In the graph of Fig. 38,

† For readers acquainted with complex numbers, this means only that there is no real solution. On graph paper we cannot show points with imaginary co-ordinates, so, for graphical purposes, only real numbers can be considered as solutions of an equation.

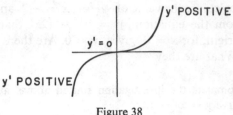

Figure 38

the curve rises at first, hesitates, and then rises again. Thus y' is positive at first, then just for a moment it is zero, then again it becomes positive. In terms of motion, such a curve might represent a person or a car moving forward, meeting some obstacle and being brought to rest, and then struggling forward again.

Exercises

1. Show that the graph of $y = x^3$ gives a curve resembling Fig. 38.

2. Sketch the graph of $y = 6x - x^2$.

3. Sketch the graph of $y = x^2 - 6x$.

4. Sketch the graph of $y = x^2 - 2x - 8$.

5. Sketch the graph of $y = x^3 - 6x^2$.

6. Are there any (real) numbers that satisfy the equation $3x^2 - 6x + 9 = 0$? Can you find any value of x for which $3x^2 - 6x + 9$ is negative? Find y' corresponding to $y = x^3 - 3x^2 + 9x$. Are there any places on the curve where y' is zero? Are there any places where y' is negative? The graph of $y = x^3 - 3x^2 + 9x$ resembles, in its main features, one of the curves shown in Figs. 30, 37, 38. Which of the three do you think it is? To check your conclusion, make a table of values of y for x from -3 through $+3$ and plot the graph by the methods you used before you met calculus.

7. Show that the curve $y = x^4 - 2x^2 + 1$ has a hilltop at the point $(0, 1)$ and valley bottoms at the points $(-1, 0)$ and $(1, 0)$. Sketch the curve.

8. A paradox—it can be shown that $y = 1/x$ has $y' = -1/x^2$. x^2 is positive whether x is positive or negative. So y' is always negative. That is to say, this curve always falls, never rises. The curve $y = 1/x$ passes through the points $(-1, -1)$ and $(2, \frac{1}{2})$ as you can easily check. If you plot these points on graph paper, you will see the second is further to the right and is higher than the first. But, if the curve is continually falling, as we move to the right we should find that we get continually lower. How, by falling steadily, have we managed to end up higher than we started? If you plot the curve carefully between $x = -1$ and $x = +2$, you should see how this strange result is to be explained.

(The result for y' above is found by writing $y = 1/x$ as $y = x^{-1}$ and using formula (4). Compare page 36, where we discussed s' for $s = 1/t$.)

The Best Way of Doing Things

Nearly every text on elementary calculus contains some examples of the following kind. "A farmer has 100 yards of fencing material. A river runs through his property. He wishes to enclose as large an area as possible, with the river as one boundary, and without buying any more material. How should he arrange his fence? The river is straight, and the area enclosed must form a rectangle." (See Fig. 39.)

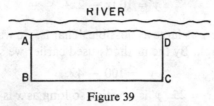

Figure 39

Whether farmers ever bother about such a problem I do not know, but problems of this type do occur in industrial design. We want the most efficient way of doing something. The actual problems may be rather complicated and may call for technical and scientific knowledge. The problem of the farmer's fence can be understood by anybody, and it should be regarded as a particularly simple example of a wide and important class of problems. It illustrates the kind of thing that can be done with calculus.

Really, the farmer has only one thing to decide—how long AB is to be. If, for instance, he decides to have AB 10 yards long, then CD also must be 10 yards long, and there will be 80 yards left for BC. The fence will then enclose 800 square yards.

There are two extreme cases of what the farmer might do. He might choose AB of zero length. Then CD would also be zero and the whole 100 yards would be available for the side BC. This would give the greatest river frontage, but the area enclosed would be zero. If he went to the other extreme, and made AD and BC both 50 yards long, he would have nothing for BC. Once again, the area enclosed would be zero. Clearly, to get the best results, he should go somewhere in between, neither try to make the enclosure as long as possible, nor as deep as possible, but somehow to balance the claims of length and breadth.

It would be quite possible to solve this problem without calculus, by drawing a graph or even simply by making a table. We would choose different values for AB, see what area each gave, and thus, by trial and error, see which was the best arrangement. If we drew a graph, we could see where the highest point of the graph came; this would represent the maximum area obtainable.

But calculus, as we have seen, gives a quick way of sketching graphs, without the trouble of making a table. So calculus gives a very neat way of solving the problem.

As we have seen, the farmer has only to choose the length of *AB*. Suppose then that *AB* is *x* yards long. *CD* also is *x* yards long. These two sides use up 2*x* yards and thus leave 100 − 2*x* yards for *BC*. The enclosure then is of length 100 − 2*x* yards and breadth *x* yards. Its area is accordingly $x(100 - 2x)$ square yards, or, multiplied out, $100x - 2x^2$ square yards. If we call the area *y* square yards, we have

$$y = 100x - 2x^2.$$

We want to make *y* as large as possible. That is, we want to choose the hilltop of this graph. By the methods used earlier we see that

$$y' = 100 - 4x.$$

So y' is zero for *x* = 25. y' is positive so long as *x* is less than 25, but negative when *x* exceeds 25 (Fig. 40).

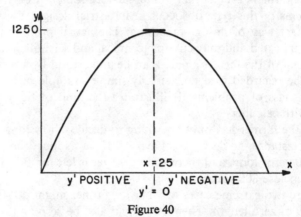

Figure 40

Thus the graph goes uphill until *x* reaches 25; it is flat for *x* = 25; and it falls once this point is passed. Clearly we have a hilltop, a maximum, where *x* = 25. Then *AB* and *CD* are each 25 yards long. *CD* is 50 yards and the area enclosed is 1250 square yards. This is the best that can be done.

As was mentioned earlier, many problems of this kind exist. Most texts on beginning calculus state and solve the well-known problem of designing a can that will hold, say, a pint of canned soup, and use as little metal as possible. Soup cans hardly ever are made in the most efficient shape; even in wartime, when it was extremely urgent not to waste metal, the less efficient shapes were still used. Some people say that it would be harder to pack and transport the cans if the design were simply based on the consideration of saving metal. Whether this is really so, or whether the soup-can manufacturers do not take calculus seriously, I have never been able to discover.

CHAPTER SEVEN

Acceleration and Curvature

If we have any expression such as $4x^3 + 5x^2$, we know how to calculate its rate of growth. We find the rate of growth is $12x^2 + 10x$. This new expression belongs to the same general type as the one we started with; someone might ask us, "At what rate does $12x^2 + 10x$ grow?" We should have no difficulty in answering $24x + 10$. The calculation is quite easy to carry out. But what is the significance of this calculation? What does the answer tell us?

We can discuss this question in terms of movement or of shapes. We will consider some examples of movement first. We begin with the law we studied in great detail earlier, $s = t^2$. For this law $s' = 2t$, where s', the rate at which s grows, is the same thing as the velocity v of the object. It does not matter whether we write $s' = 2t$ or $v = 2t$. Now, we ask, "How fast is $2t$ growing?" This could equally well be phrased, "How fast is v growing?" The natural symbol for the answer is accordingly v', the rate at which v grows. As $2t$ grows at the rate 2, we have $v' = 2$. Putting all this together, we have

$$s = t^2,$$
$$v = s' = 2t,$$
$$v' = 2.$$

This last equation, $v' = 2$, tells us how fast the velocity is increasing. The rate of increase of velocity is usually called the acceleration. Acceleration is usually denoted by the symbol a; some books use f. We shall use a.

71

Now we have three things to bear in mind, distance, velocity, and acceleration; in any statement we read, we must be careful to see whether it is a statement about s, v or a.

In a car, where would you look if you wanted to know the value of s? You would look at the mileage recorder, or at the milestones by the side of the road. s tells you how far you have come. How would you tell the value of v? The simplest way would be to look at the speedometer. If that was out of action, you might look at the mileage recorder and see *how fast* the numbers were clicking over, or you might look out of the window and see how quickly the milestones were whizzing past. The speedometer gives you directly the value of v (unless it is out of order); the other methods depend on your estimating s', the *rate* at which your distance from home is increasing. Now, what about the acceleration a? Where would you look for that? So far as I know, no car has a dial to tell the driver the value of a. But as $a = v'$, the rate of growth of velocity, we can estimate a by looking at the speedometer needle and seeing *how fast it is moving*. We do *not* find a simply by reading the speedometer. A car traveling at a steady 100 mph is going quite fast; nevertheless, its acceleration is zero. The needle is at rest at the mark 100 mph. On the other hand, a car can be moving quite slowly and yet have a large acceleration a. If your car is starting from rest, at first the speedometer needle points to 0 mph; soon after it points to 5 mph, then to 10 mph, and so on. The velocity is small but it is growing. If you are one of those people who accelerate fiercely from rest, the car's velocity might climb from 5 mph to 10 mph in a very short time; in that case the acceleration a might be quite large although the velocity is still quite small.

There is another way of estimating acceleration. When a car accelerates hard, the passengers tend to be thrown back in their seats. In the same way, if the driver suddenly jams on the brakes the passengers tend to fly through the windshield. Braking is negative acceleration. So acceleration is something you can feel. When a car accelerates positively, you can feel your seat pressing you forward. If it accelerates negatively, you may feel the windshield pushing your head back. Acceleration is what hurts. It does not hurt you to travel at 200 mph. Plenty of people have done that in an airplane. It is quite all right so long as you know that you have plenty of room to keep moving. What hurts is if you hit a wall when you are traveling at 200 mph. Then you are suddenly brought to rest; you have a large negative acceleration. It is just as bad for someone sitting on the wall when you hit him. He is at rest and suddenly you try to make him travel at 200 mph like yourself. He experiences a sharp positive acceleration. This is just as painful for him. It is the same if I suddenly give you a vicious kick.

You, or the part of you that gets kicked, suddenly experiences a large acceleration; it may hurt my foot too, because my foot is suddenly brought to rest. Large accelerations mean large forces. It is not surprising that in mechanics the force acting on a body is not measured by the position s of the body, nor by its velocity v, but by its acceleration a. The Earth rushes along its orbit round the Sun at about 1000 miles a minute, but we do not feel this at all. We should feel it indeed, if we banged into some object that was at rest relative to the Sun and we were suddenly brought down from 1000 miles a minute to a much smaller speed.

It is good to think of all kinds of situations, and see how they would be described in terms of s, v, and a. For example:

1. *A breakdown.* The car is parked at the side of the road. The mileage recorded does not change. That is, s is constant. The velocity is zero. So is the acceleration. In equations,

$$s = c, \text{ a constant,}$$
$$v = 0,$$
$$a = 0.$$

2. *Touring at a steady speed.* The car is traveling at 60 mph on a straight, deserted parkway. The law could be $s = 60t$. (It could also be other things, for example, $60t + 100$ or $60t - 30$, depending on what instant we measure our time from.) The velocity is 60. Since the velocity is steady, there is no acceleration.

$$s = 60t,$$
$$v = 60,$$
$$a = 0.$$

Note that each expression gives the rate of growth of the one above it, and we are reminded again that the rate of growth of a constant is zero.

3. *Accelerated motion.* A car is moving forward with increasing speed. I do not want to go into questions of engine performance, so I shall just use our stock example for an object gathering speed, $s = t^2$. Then we have

$$s = t^2,$$
$$v = 2t,$$
$$a = 2.$$

The acceleration here is constant, which means that the car is being driven forward by a constant force. I very much doubt if an internal combustion engine would behave in this way. Perhaps we had better suppose that the car is in neutral, and is just being allowed to roll forward on a gentle slope. And I suppose, in this example, we had better use feet and seconds as units. It is most inconvenient to measure acceleration in "mph an hour".

4. *Braked motion*. A car is drawing smoothly to a stop. There are many laws that would fit this situation. I choose the simplest one I can think of that will give the required type of motion, say, $s = 10t - t^2$ between the times $t = 0$ and $t = 5$. (The work below shows that this does in fact represent braked motion.) By considering rates of change, we get the following:

$$s = 10t - t^2,$$
$$v = 10 - 2t,$$
$$a = -2.$$

I think the middle equation gives the clearest picture of what is happening. At the beginning, $t = 0$ and $v = 10$. So to begin with, the car is moving at 10 feet a second. Five seconds later, $t = 5$ and so $v = 0$; the car has come to rest. If you calculate the velocity v at times between you will obtain the following table:

t	0	1	2	3	4	5	
v	10	8	6	4	2	0	.

So the car is losing speed in a perfectly regular way; v decreases by 2 with every second that passes. And that is precisely what the third equation, $a = -2$, tells us. What distance does the car travel while coming to rest? For this information we must turn to the first equation. When $t = 0$, $s = 0$. When $t = 5$, $s = 25$. So the car advances 25 feet while coming to rest; this is very gentle braking.

If we were to put $t = 6$ in the equation above, we should find that $v = -2$; that is to say, the car had started to move in reverse! Of course, this would be an incorrect conclusion to draw. Brakes make a car slow down so long as it is advancing, but they do not cause it to retreat after it has come to rest. The law $s = 10t - t^2$ applies only during the breaking, from $t = 0$ to $t = 5$. We have no right to assume that it applies either before $t = 0$ or after $t = 5$.

One can however imagine circumstances in which this law would apply after $t = 5$ as well as before. Suppose, instead of the brakes being on, the driver sees that the road in front of him is going uphill. To save his brakes, he decides to let the hill bring the car to rest; he puts in the clutch and waits for the car to slow down under the influence of gravity. If he does not remember to put the brake on as soon as the car has come to rest, the car will begin to roll back down the hill; if the driver allows it, the car may return to the place where it was when we started to observe it. In that case, the law $s = 10t - t^2$ might apply to the car between $t = 0$ and $t = 10$. The following table shows the position, velocity, and acceleration of the car throughout this process.

Time, t	0	1	2	3	4	5	6	7	8	9	10
Position, s	0	9	16	21	24	25	24	21	16	9	0
Velocity, v	10	8	6	4	2	0	−2	−4	−6	−8	−10
Acceleration, a	−2	−2	−2	−2	−2	−2	−2	−2	−2	−2	−2

Each row of the table gives its own kind of information. s shows you where the car is at any time. You notice that it ends up where it started. v shows how fast it is going. To begin with, it is advancing at 10 feet a second; after 5 seconds, it has come to rest; at the end, it is rolling backwards just as fast as it was rolling forwards to begin with. The third row a contains the same number −2 throughout; this means that gravity is always dragging the car back with the same force.

Above we have used the letters s, v, a and have had to remember that these measured distance, velocity, and acceleration. Calculus notation brings out the fact that v tells us how quickly s grows, and a tells us how quickly v grows. This, as we have seen, can be written $v = s'$ and $a = v'$. In the equation $a = v'$, we could make use of the fact that v is equal to s'. If we substitute s' for v, we get $a = s''$. In words, this last equation states that a gives the rate of change of the rate of change of s. In future, as a rule, instead of using s, v, a for distance, velocity, and acceleration, we shall use s, s', s''. In the same way, when we are dealing with graphs, we shall meet the symbols y, y', y''. y' tells us how quickly y grows; y'' tells us how quickly y' grows. When we are dealing with particular formulas, the procedure for finding y' and y'' is very simple. For example, suppose $y = x^5$. What is y'? We know from our earlier work that x^5 grows at the rate $5x^4$. So $y' = 5x^4$. Now what is y''? y'' is the rate at which y' grows. y' is $5x^4$. We know that $5x^4$ grows at the rate $20x^3$. So $y'' = 20x^3$. It is no harder to get from y' to y'' than it was to get from y to y'—that is, so far as making calculations is concerned. We still have to interpret these calculations to see the meaning of y'' for a graph. That is our next job. We begin by collecting together the four examples of motion we considered above. We describe each type of motion by means of s, s', s''; we also describe it in words; we give the corresponding graph, and we repeat our information about s, s', s'' in symbols suitable for a graph, that is, in terms of y, y', y''. The whole thing now looks like this:

Type of motion	Distance, s Velocity, s' Acceleration, s''	Graph	Graph in terms of y, y', y''
Object at rest	$s = c$ $s' = 0$ $s'' = 0$	——— Figure 41	$y = c$ $y' = 0$ $y'' = 0$
Motion at steady speed	$s = 60t$ $s' = 60$ $s'' = 0$	Figure 42	$y = 60x$ $y' = 60$ $y'' = 0$
Accelerated motion	$s = t^2$ $s' = 2t$ $s'' = 2$	Figure 43	$y = x^2$ $y' = 2x$ $y'' = 2$
Braked motion	$s = 10t - t^2$ $s' = 10 - 2t$ $s'' = -2$	Figure 44	$y = 10x - x^2$ $y' = 10 - 2x$ $y'' = -2$

We already know what y' tells us; it gives information about the steepness, the slope of the graph. We want to know what y'' signifies. First of all, what it does *not* signify. Students sometimes get confused and say, "When y'' is zero, the curve is flat." This is certainly not so. For the first two graphs above, Figs. 41 and 42, the value of y'' is zero throughout. Admittedly, Fig. 41 shows a curve that is flat throughout. But Fig. 42 also has y'' zero, and this certainly represents a curve that is climbing, not one that is flat. $y'' = 0$ both for Fig. 41 and for Fig. 42, so $y'' = 0$ must represent a property that is *common* to Figs. 41 and 42.

INVESTIGATION. Sketch a number of graphs such as the following: $y = x$; $y = 2x$; $y = 2x + 3$; $y = 6 - 2x$; $y = -x$; $y = x + 2x^2$; $y = x + x^2$; $y = x - x^2$; $y = x - 2x^2$. Work out y'' for each of these. Collect your graphs into the following three groupings:

Type A—graphs for which $y'' = 0$ throughout.

Type B—graphs for which y'' has a positive value throughout.

Type C—graphs for which y'' has a negative value throughout.

All the graphs of type A have a certain property which distinguishes them from those in type B and type C. What is this property? In the same way, all the graphs of type B have a distinctive property in common; what is it? And again, what is the property that gives a family resemblance to all the graphs in Type C? If you can answer these questions, you will be able to tell just by looking at a graph whether it belongs to type A, B or C. If you find the examples given above are insufficient and you would like to have some more material to examine before reaching a decision, sketch further graphs of linear and quadratic expressions. That is, choose equations of the form $y = mx + k$ or $y = ax^2 + bx + c$. With equations of higher degree you may get graphs that do not fall into any of the types A, B, C.

Classify the graphs shown in Fig. 45 as being in type A, B or C.

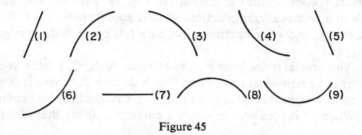

Figure 45

If you possibly can, complete this investigation before reading on.

*   *   *

By looking at graphs of type A, B, C, you should have obtained, at the very least, a strong suspicion of what the sign of y'' means. We shall now approach the same question in a different way.

The sign ' stands for the "rate of growth of." If z stands for any quantity whatever, z' stands for the rate of growth of z. If z' is positive, this means that z is increasing; that it is changing by having something added to it; that it is actually *growing* in the everyday sense of the word. If z' is negative, this means that z is growing in a negative sense; that it is changing by having something subtracted from it; that it is decreasing or shrinking. Now y'' measures the rate of growth of y'. If y' is increasing, y'' will be positive; if y' is decreasing, y'' will be negative.

The curve in Fig. 46 is flat at first, and at the end has a direction pointing to the north-east. In numerical terms, it begins with $y' = 0$ and ends with $y' = 1$. So y' is increasing and y'' is positive.

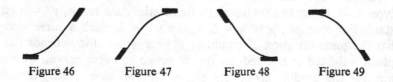

Figure 46   Figure 47   Figure 48   Figure 49

The opposite happens for the curve in Fig. 47. This begins pointing

north-east, and ends flat. y' is 1 at first and 0 at the end. y' has de-
creased. Its rate of growth is negative, so y'' is negative.

We need to be careful with the two remaining examples. In Fig. 48
the curve starts off in a south-easterly direction and ends flat. South-
east corresponds to the slope $y' = -1$. So in this curve, y' changes
from -1 to 0. Is this an increase or a decrease? We have to add 1 to
-1 to obtain 0. From -1 to 0 is an increase (think of temperature,
for example). So y' is increasing and y'' is *positive*. Compare this with
your graphs of type B and verify that it does resemble them.

Finally, consider the curve shown in Fig. 49. This starts off flat and
ends in a south-easterly direction. y' thus goes from 0 to -1. So y' is
decreasing, and y'' is negative. Compare this curve with your graphs
of type C.

If you now examine your graphs of types A, B, C, I think you will
see what I mean when I say that y'' tells us how the curve is *bending*.
When y'' is zero throughout, we obtain a straight line, no bending at
all. When y'' is positive, we obtain a curve resembling that of a plank

Figure 50

with a weight in the middle (Fig. 50). When y'' is negative, we obtain
a curve resembling a plank with weights on the ends (Fig. 51). In fact,
in reinforced concrete design and in other branches of civil engineering,
y'' appears in just this role, as a measure of bending.

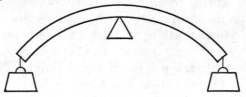

Figure 51

All the equations that we used when we were drawing graphs of
type A, B, C were either linear or quadratic. As a result, y'' was constant. For example, $y = x + x^2$ gave $y'' = 2$. Such a curve always
has the same direction of bending; $y = x + x^2$ always bends like a
plank weighted at the middle. But if we go to cubic curves, we need
not find the bending always the same. This we have already met. On
pages 65–67, we considered the graph of $y = x^3 - 12x$. The final
sketching of this curve was left as an example for you. By considering

$y' = 3x^2 - 12$, we saw that this graph rises until $x = -2$, sinks between $x = -2$ and $x = 2$, and then rises again. In fact, it appears as in Fig. 52. What about the bending of this curve? If you looked at the part of the curve to the left of the origin, you might well think this was a parabola of type C; if you looked at the curve to the right of the origin, you might well think this was a parabola of type B. Do not

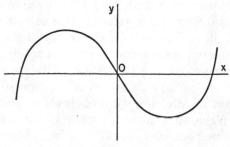

Figure 52

mistake me; this curve does not in fact consist of two parabolas fitted together. It merely has a very general resemblance to what you would get by joining two parabolas. To the left of the origin, we see the type of bending ("plank with ends loaded") that we associate with y'' negative; to the right we see the type of bending we associate with y'' positive ("plank with middle loaded"). How does this agree with the information we obtain from the equation of the curve? From $y = x^3 - 12x$, we find $y' = 3x^2 - 12$ and $y'' = 6x$. Now $6x$ is negative when x is negative and positive when x is positive; so y'' is negative to the left of the origin, and positive to the right. This fits in exactly with the type of bending we have observed.

In Fig. 33, we sketched the graph of $y = 100x - x^2$ by studying the behavior of y'. We saw that the graph rose until $x = 50$ and fell after that. Then the question was raised, how do we know that the graph does not have a whole lot of little wobbles in it, like the curve in Fig. 34? We can now answer this question. Since $y = 100x - x^2$, we have $y' = 100 - 2x$ and $y'' = -2$. So y'' is always negative, whatever x. That means the bending is always of the same kind, like the plank. This cuts out the possibility of wobbles, for wobbles mean that the curve bends first one way, then the other.

In our discussion, we have only considered the sign of y''; we see where y'' is positive, where negative, where zero. It is possible, by considering the actual sizes of y' and y'' to find out not only which way, but how fast, the curve is bending. We can say that, at a particular point, a curve is bending in the same way as a circle of radius 3, say. At another point, where the curve has a hairpin bend, it might be bending like a circle of radius 0.1.

There is a branch of mathematics known as differential geometry. It applies the methods of calculus to the study of geometrical objects, such as curves and surfaces. The question just mentioned, of finding how quickly a curve is bending, belongs to the province of differential geometry, and is a very simple example of a problem in that subject. Differential geometry also deals with the curvature of surfaces. The study of curved surfaces leads naturally to a subject known as tensor calculus, which in turn is used for the theory of relativity. You may have heard rather mysterious references to "curved space-time." This is a good example of the way in which calculus opens the door to all kinds of investigations. You start off with some simple devices for sketching graphs rapidly; one question leads to another; you study curves in the plane, then curves in space of three dimensions, then surfaces; new methods of calculation, new symbols, new concepts gradually creep in; you end up, in a way that could never have been foreseen at the beginning, with a theory that has revolutionized our ideas of space and time, of gravitation and energy.

CHAPTER EIGHT

The Reverse Problem

In elementary arithmetic there are certain simple, direct processes, to add, to multiply, to square. These can always be carried out within the framework of the natural numbers, the counting numbers 0, 1, 2, 3, 4, ... What is 3 added to 4? Answer: 7. What is 3 multiplied by 4? Answer: 12. What is the square of 3? Answer: 9.

Later on, we learn to reverse these operations. We learn to subtract by reversing addition. 3 and what make 7? Answer: 4. We divide by reversing multiplication. 3 times what is 12? Answer: 4. We reverse squaring to extract square root. Which number squared gives 9? Answer: 3.

These reverse operations lead us to extend our ideas. When we try to answer the question, "8 and what make 7?", we may at first say there is no answer; later, we discover that one can introduce a new idea, that of negative numbers, and then we can answer, -1. In the same way, division leads to an idea that was at one time new, the idea of fractions. Instead of saying that "2 times what is 1?" has no answer, we arrive at the answer 1/2. Square root leads to yet new ideas; we cannot write down any fraction (in the sense of elementary arithmetic) whose square is 2. We are thus led to irrational numbers such as $\sqrt{2}$. If we look for a number whose square is -1, we are led to the even more striking idea of complex numbers such as $\sqrt{-1}$.

In calculus, exactly the same kind of growth occurs. We began with the direct question, "I give you a law which tells you where an object is at any time. You are to give me a law for its velocity." We can easily reverse this; I give you a law for the velocity; you are to give me a law for the position. In symbols, I give you a law for s' and request a law for s. Sometimes this question is easily answered. For example, if I give the law $s' = 2t$, you might answer with the law $s = t^2$ or $s = t^2 + 5$ or $s = t^2 - 3$ or in fact with any formula of the type $s = t^2 + C$, where C is a constant. But such questions can lead to new types of formula. For example, I might give the law $s' = 1/t$ and demand a law for s; to answer this question, you have to develop the theory of logarithms. If I give the law $s' = 1/\sqrt{1 - t^2}$, to find a law for s you have to develop the theory of the trigonometric functions, sine and cosine. In high-school work, trigonometry is usually thought of as something to do with surveying. It is approached through the geometry of triangles. The calculus approach is quite different. There is no mention either of geometry or of surveying in the question of finding s given $s' = 1/\sqrt{1 - t^2}$. Calculus thus brings us into trigonometry by what might be called an algebraic approach; I mean by this that we are writing equations, not drawing pictures. The calculus approach helps to pull our mathematics together. Trigonometry does not appear as a separate subject, but arises quite naturally in the development of calculus. Also calculus gives us some information about trigonometry that would be very hard to obtain without calculus. Students sometimes ask, "How are the trig tables calculated?" The answer is to be found in calculus.

Trigonometry is only one of the subjects that arises in this way. As we continue to study the problem of finding s when s' is given, we are led to study new types of function which do not occur at all in high-school mathematics.

We are also led to new types of function in another way. In algebra we can form equations. We are not confined to simple processes like extracting square root. We may ask, for example, for a number whose square exceeds the number itself by 20. In symbols, we have to solve the equation

$$x^2 = x + 20$$

and this of course is very easily done. Other equations are not so simple to solve. For instance, mathematicians spent several centuries before they knew all about equations of the type

$$x^5 = x + 20.$$

In calculus we can also form equations. One might ask whether there is any law for s such that

$$s' = \frac{2s}{t}.$$

This question is easy to answer. The law $s = t^2$ would be a solution. For, if $s = t^2$ we have $s' = 2t$. Then $2s/t = 2t^2/t = 2t = s'$. So the law $s = t^2$ has the property desired. There are many other solutions of this question. $s = 5t^2$ or $s = 7t^2$ would do just as well; in fact any law of the form $s = kt^2$, where k is constant, will do.

The equation above can be put into words. s' is the velocity at any time. s/t, being the total distance gone divided by the total time taken, measures the average velocity. So, the equation asks: "Can you find a type of motion in which the velocity, at any moment, is exactly double the average velocity for the journey up to that moment?" The answer is that our familiar motion with constant acceleration, $s = kt^2$, will do the trick.

You may wonder why I chose this particular problem. The answer is simple. I did not want to get involved in long and difficult calculations, so I looked for a problem with an easy answer. In fact I started with the solution $s = t^2$ and worked backwards; I looked for an equation that would have this as a solution.

Our problem could be put geometrically. In terms of x and y, the corresponding equation would be

$$y' = \frac{2y}{x}.$$

Consider Fig. 53. Suppose P is the point (x, y) on the curve AB. PT is the tangent at P. OPR is the straight line joining the origin O to the point P. PQ is a horizontal line and QRT is vertical. We can now interpret our equation geometrically. y' of course gives the steepness of the tangent PT. y/x gives the steepness of the line OP. The equation requires that y' be just double the size of y/x, that is to say, the slope of the tangent PT is to be exactly twice the slope of the chord OP. This

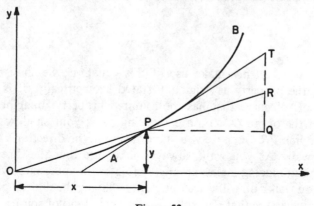

Figure 53

means that QT would have to be exactly twice as long as QR. This property has to hold for every point of the curve AB. So the problem is to find a curve AB such that for any point P of the curve, the slope of the tangent PT is exactly double the slope of the line OP. The solution is that any parabola of the form $y = kx^2$ has this property. Without calculus this would be an extremely awkward problem to solve. This problem is, of course, of no great significance; it was selected because it was easy to solve. But problems similar to this do arise in the course of actual investigations in engineering, science and pure mathematics.

The problem above had a solution in terms of simple, known laws. No new idea enters into the equations $s = kt^2$ or $y = kx^2$. In the graphical problem we recognize the curve as a parabola. Now there are fairly few curves that we know by name; the straight line, the circle, the ellipse, the parabola, the hyperbola—this is the whole list for many people. Even if you are particularly interested in curves, it is unlikely that you know more than twenty by name. There are thousands that you do not know and would not recognize. It is therefore unlikely that a problem will lead to a curve we already know. The chances are overwhelming that it will lead to a curve we do not yet know.

This sounds rather disturbing. It looks as if we are going to be pretty helpless in the face of problems. However it is not as bad as it sounds. It is true that most problems lead to new curves. However, the problem itself tells us what the new curve is going to be. In effect, the problem defines its own solution. This may be seen by considering an example.

Suppose we were asked to find a curve with the property that, for each point P, the tangent PT makes an angle of $45°$ with the line OPR, as shown in Fig. 54. I need not bother you with the details of the calculation, but this property would be expressed by the equation

$$y' = \frac{x + y}{x - y}.$$

However we shall not make use of this equation; we shall think in terms of the property as originally stated geometrically. It is easy to see the sort of curve that has the required property. Imagine a light placed at the origin O. You are standing at P; your shadow will fall along the line PR. Suppose you stand facing in the direction PT. Now start to walk. As you walk, always make sure that the direction in which you are facing makes an angle of $45°$ with your shadow. In this way as you walk you will trace out a curve having the desired property. I think you can see that you will describe some kind of spiral; you will

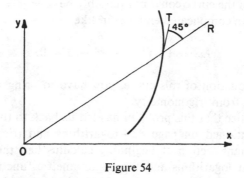

Figure 54

keep going round the light, but always getting further away from it. It would also be possible to produce this curve by a mechanical contrivance (see Fig. 55). Here, *OR* is some kind of rod or bar. A nail fastens the point *O* to the paper, but the rod *OR* can revolve about *O*. At *P* we have a little sleeve, which can slide freely on the rod. Underneath this sleeve there is a little wheel with a sharp edge, so fixed that it always makes an angle of 45° with the rod *OR*. The edge of this wheel cuts into the paper, and so forces *P* to move only in the direction *PT*. If the rod *OR* is now turned, *P* will automatically move in the required way. It will trace out the curve in the same way as you did by walking according to the specifications given above.

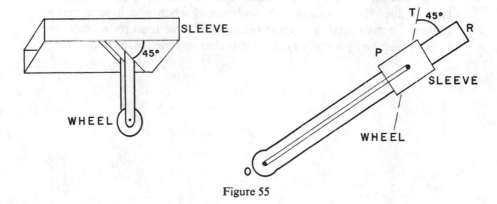

Figure 55

As you will see, here the problem itself has shown us how to construct the curve. The question really is, that we should find *some other way* of specifying the curve. We might find the equation of its graph; this might give us a more convenient way of defining this curve. This particular question has been investigated, and it has been discovered that the curve cannot be specified by elementary algebraic operations.

The equation of the curve could not possibly be, say, $y = x^3 + 5x - 2$, nor even a more complicated expression like

$$y^5x^3 - 17xy^2 + 11x^4 - 3 = 0.$$

To write the equation of this curve, you have to bring in ideas from logarithms and from trigonometry.

You may notice that this problem has led us back to the same topics that were mentioned on page 82—logarithms and trigonometry. In fact, an important section of beginning calculus had the two aims— to explain what logarithms and the trigonometric functions are, and then to show what problems can be solved with their aid.

I hope the purpose of this chapter, under the heading "The reverse problem", will be clear to you. Its purpose is not to teach you any particular result. Rather, the object is to give you some idea of mathematics as a growing subject. We began by studying the speed of a moving object; the few formulas and the symbols s', s'' which we met now allow us to pose new problems. Some of the problems lead to branches of mathematics, the names of which you know, such as trigonometry and the theory of logarithms. There are other problems which lead on to branches of mathematics of which you have never heard even the names. The ideas of elementary calculus, as I mentioned earlier, are in fact the key that opens the door to most of the mathematics and most of the science developed between 1600 and 1900 A.D. How it does that, you can only understand when you have actually studied the mathematics of these centuries. I have tried to indicate, in a very vague and general way, how it is that one idea leads to another.

CHAPTER NINE

Circles and Spheres, Squares and Cubes

We have so far used only the letters s, t, and x, y in our work. In algebra, of course, one letter is as good as another. We can pass from $s = t^2$ to $s' = 2t$, and from $y = x^2$ to $y' = 2x$; in the same way, we can pass from $p = q^2$ to $p' = 2q$ or $J = w^2$ to $J' = 2w$.

When you were young, you met the formulas for the area of a circle, $A = \pi r^2$ and the circumference $C = 2\pi r$, also for the volume of a sphere, $V = (4/3)\pi r^3$, and the surface of a sphere $S = 4\pi r^2$. Now that you have studied calculus, something may strike you about these formulas. Suppose you take $A = \pi r^2$ and ask, what is A'? r^2 grows at the rate $2r$. What shall we do about the coefficient π? π, of course, is a fixed number, although it is written in this strange way with a Greek letter. If we had $A = 3r^2$, we should go easily enough to $A' = 6r$ (see formula (6) and the picture of the growing plant, Fig. 18). π is just a bit more than 3 and we treat it in the same way. From $A = \pi r^2$, we find $A' = 2\pi r$. But we recognize this result; $2\pi r$ gives the circumference of the circle. So $A' = C$.

We find a very similar result for the sphere. From $V = (4/3)\pi r^3$ we obtain $V' = 4\pi r^2$, so $V' = S$. This can hardly be a coincidence. In fact it is easy to see why it occurs. Suppose you have a sphere and you want to make it a little larger. You might spray an even coating of plastic all over its surface, thus giving it an extra skin. It is not at all surprising that the amount the volume has increased during this operation should be closely related to the area of the surface on which the skin has been placed.

One danger point in this argument is to be noticed. It is absolutely essential that the coating should be even; the skin must have the same thickness everywhere.† One can reach very strange results by applying this argument to an egg-shaped solid. For if you enlarge the scale of an egg, you do *not* add a layer of uniform thickness.

Areas and Volumes

The idea that objects grow by forming an extra skin can be illustrated without using circles and spheres. Two of our first calculus results can be illustrated by means of squares and cubes.

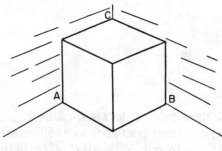

Figure 56

Imagine a cube placed in the corner of a room (see Fig. 56). This cube is growing, because someone is continually spraying plastic onto the exposed faces of the cube. This is done in such a way that the points A, B, C move outwards at 1 inch a second. If we begin with no cube at all at $t = 0$, and let it grow from nothing in the way specified, after t seconds the side of the cube will be t inches. Its volume will be $V = t^3$. This, we know, grows at the rate $V' = 3t^2$. The picture shows why $3t^2$ should come into it, for the exposed surface consists of three squares, each of area t^2; each surface is moving outward at unit rate, so the surface area $3t^2$ gives the rate at which the cube is growing fresh skin at any moment.

The cube is a figure in three dimensions. A similar thing happens in two dimensions. Imagine you are given the following job. Figure 57 shows two pointers that move along the lines OX and OY at unit speed. You have a pencil, and you must keep shading the paper so that there

---

† In effect, we are estimating the increase in the volume by multiplying the surface area by the thickness of the skin. This estimate is "reasonable" if the coating is thin. One has to think very carefully to show that the argument is in fact logical and gives the exact result.

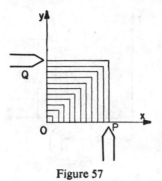

Figure 57

is always a square shaded. You have to keep pace with the moving pointers; your square must always reach just to the pointers P and Q. At first you have a fairly easy job, but as the size of the square grows, you have to draw longer and longer lines. At time t, the square will have side of length t inches. The area will be $A = t^2$, and $A' = 2t$. Your pencil line runs round two sides, so the boundary has length $2t$. Here again the length of the boundary agrees with the value of A'.

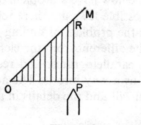

Figure 58

Suppose we chop our last figure in two by a line OM, the bisector of angle YOX. The lower half will be as shown in Fig. 58. The shaded area is now $\frac{1}{2}t^2$, growing at the rate t as the pointer P moves outward. If we keep shading in the area of the triangle, we shall at any moment be drawing a line such as PR, of length t. If we now denote the area of the shaded triangle by A, we shall have $A = \frac{1}{2}t^2$ and $A' = t$. A' agrees with the length of the boundary PR.

This may make us wonder: does the line OM have to be straight? Could we not consider a picture like that of Fig. 59? Here OM is the parabola $y = x^2$. Once more, P moves at unit speed, and the shaded area grows by developing a skin along the line PR. At time t, the distance OP is t. The x co-ordinate of the point R is thus t. Since R lies on the graph $y = x^2$, the point R must have $y = t^2$. Thus the line PR is of length t^2.

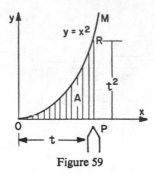

Figure 59

On the analogy of our previous results, we are led to guess that the area A will grow at a rate equal to the length of PR. That is, we expect to find $A' = t^2$. This guess is in fact correct. The shaded area is given by the law $A = \frac{1}{3}t^3$ and this does make $A' = t^2$.

I am not trying here to discuss the theory of area in any detail. I am trying to show how the idea of speed links onto that of area. In many cases, when we are trying to determine an unknown area A, we find that it is possible to calculate *the speed at which this area grows*. That is, we can find A' and we now have a problem of the reverse type; given the law for A', what possible laws are there for A?

You will notice that the problem of finding the area underneath the parabola $y = x^2$ is quite different from the elementary problems about the areas of triangles, parallelograms, and rectangles. It looks much harder. Calculus gives us a way into it. The actual calculation proves to be quite simple. You will find the details in any introductory text on calculus.

Finding volumes is much like finding areas. You are doubtless familiar with the formula $V = (4/3)\pi r^3$ for the volume of a sphere of radius r. This formula was mentioned in the previous section of this book. But, although you know this formula, it is unlikely that you know how it is obtained; to find the volume of a sphere is in fact a calculus problem. It may interest you to have a brief sketch of the idea used.

We will consider the volume of a sphere of radius one inch. In Fig. 60, the circle is supposed to be of radius one inch and centre O. We cannot show a sphere properly on paper. You must imagine the picture spinning around the line OX. The circle will then cut out a sphere in space. It is this sphere we have to think about. You can imagine it as a hollow metal sphere. We are going to fill it up soon. The line DOC, as it spins about OX, will sweep out a circular disc. Imagine this disc as being a piece of paper that divides the interior of the sphere into two parts. We now start to fill the sphere up. We might bring along a series of circular discs of paper, and keep pasting these on. The discs,

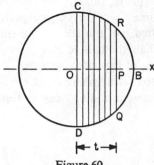

Figure 60

of course, would not all have the same radius. If we had at some stage
filled the shaded region, the next disc we pasted on would have to have
the radius *PR*. This disc would occupy the region of space obtained by
spinning the line *QRP* around *OX*. Or, instead of discs, we might
imagine plastic sprayed on in layers. Either way, we imagine the shaded
region to grow out towards the right, so that the distance *OP* grows at
unit rate. Thus after *t* seconds, the distance *OP* would be *t* inches. At
any stage of the process, the region filled will be that part of the sphere
lying between two parallel planes.

Every new skin that comes on is even; it has the same thickness at
all points. Further, the surface is moving outward at unit rate. So, as
in our earlier examples, the area of the surface being coated gives us
the rate of growth of the volume. What is the area of this surface? The
surface is a circle of radius *PR*. We must calculate *PR*. This is not
difficult. *OPR* is a right triangle. *OR* = 1 since the circle is of radius
1 inch. *OP* = *t*, as we noted in the previous paragraph. By the Pytha-
gorean Theorem, we find $PR^2 = 1 - t^2$. Fortunately it is the square
of *PR*, and not *PR* itself, that we need. The area of the circle of radius
PR is $\pi \cdot PR^2$. This is $\pi(1 - t^2)$. If *V* stands for the volume that has
been filled by time *t*, we have

$$V' = \pi(1 - t^2).$$

Here again we have a reverse problem. We know how fast *V* grows;
we know that *V* begins by being zero at *t* = 0. This information is
sufficient to give us the answer,

$$V = \pi(t - \tfrac{1}{3}t^3).$$

You may notice that we have arrived at the answer to a harder
problem than the one we set out to solve; this formula gives us, not the

volume of the whole sphere, but the volume lying between two planes. However, when the time $t = 1$ arrives, the point P will have reached B, and we shall have filled half the sphere. Putting $t = 1$ in the formula above, we find that half the sphere has the volume $\pi(1 - \frac{1}{3})$, that is, $\frac{2}{3}\pi$. To find the volume of the whole sphere, we double this, and obtain the answer we expected for a sphere of unit radius, namely $(4/3)\pi$.

The volume of a sphere of radius r can be found by the same method.

CHAPTER TEN

Intuition and Logic

You have now had some samples of the problems of beginning calculus and some indications of the questions that cause calculus to develop beyond these. With the background given here, there are many texts on calculus that you should be able to read for yourself and understand without difficulty. There will be others that do not seem to make sense to you at all. And some books will lie halfway between these. You will be reading quite happily and then you will come to a page that seems to you entirely unnecessary. Perhaps you will not understand what it is saying at all; again, you may find that long arguments are used to reach a conclusion that seems perfectly obvious.

To understand this, you need to know something of the history of mathematics. During the years 1600–1800 A.D., calculus was concerned with very much the kind of problems, and used very much the kind of thinking, that you have seen in this book. Then, gradually, a crisis developed. As mathematicians explored deeper and deeper into the subject and studied more and more complicated situations, they began to get answers that were evidently wrong. Their way of thinking, which had been perfectly satisfactory for dealing with simpler situations, was now proving unreliable; they found it necessary to examine very carefully things which before they had taken for granted.

Such a crisis is nothing unusual, and nothing to be ashamed of. In fact, a crisis is often a sign of health. A growing boy finds he cannot wear his old clothes; he is too large for them; he needs new ones. In the same way, a growing subject from time to time needs new ways of thinking; it grows out of the old ones.

This, of course, creates a problem for teachers. Shall we give beginners in calculus the suit of ideas that fitted the mathematician of the year 1700? If we do, there is the danger that he may become accustomed to this suit, and refuse to change into more grown-up garments later on. On the other hand, if we give the student the 1961 model, he may find that the arms and legs are much too long for him, and prevent his moving about at all.

Mathematicians differ among themselves on the correct answer to these questions, and of course students differ too; one student may enjoy a type of teaching that is utterly distasteful to another. My own belief is that, for the majority of students, it is unwise to begin with the latest and most fashionable model. It is better to begin with a more modest but better fitting suit. But you should realize that you will not wear this all your life.

What then are these different ways of thinking? If you look back through this book, you will find that a large number of ideas have been taken from everyday life—moving bodies, velocity, acceleration, slope, area, volume. We have not tried to give exact definitions of these ideas; we have assumed that we understood more or less what these words meant, and we have argued on that basis.

Mathematicians call this the *intuitive* approach. In daily life, our thinking is nearly always intuitive. Few of us could give an exact definition of the word *dog*. But we recognize dogs when we meet them. There may be some doubtful borderlines—just when does an animal cease to be a dog and become a wolf?—but we do not trouble ourselves too much about these. And with this kind of thinking we do manage quite well in practice, so there must be something quite sound in it.

During the 17th and 18th centuries, as was mentioned earlier, mathematicians were very much concerned with scientific problems. They wanted to determine in which curve the earth moved round the sun, and how its velocity varied as it went around. They did not see much point in philosophical discussions about *velocity*. They were sure the earth had a velocity and they wanted a formula for it.

Intuitive thinking, as you see, mixes together mathematics and physics. In this book, we have frequently used such a mixture. Our line of argument has been something like this:

(A) The wagon in Fig. 13 on page 23 moves according to the law $s = t^2$.

(B) At any moment, this wagon has a velocity.

(C) We want to find out what that velocity is.

We did in fact succeed in finding the formula $s' = 2t$ for the velocity.

Look at statement (B) above. This statement seems to imply that, when a body is moving, it must have a velocity; it must be moving at some speed.

It is very natural to assume this. Most people, if they were told something was moving, would ask whether it was moving fast or slow. They would be most astonished if they were told that the object was moving, but was not moving at any speed.

The earlier mathematicians did not even consider such a possibility. However, in the 19th century, mathematicians found themselves confronted with actual examples of formulas where this happened—a point was moving, but not moving at any speed!

We can state this apparent paradox another way. From the beginning of this book, we have used pictures to show the motion of an object. In these pictures, the velocity of the object corresponds to the steepness of a curve. Thus the direction of the curve corresponds to the velocity of the object. If the object does not have a velocity, this means that the corresponding curve *does not have a direction*!

You may find it very hard to imagine such a thing existing. If so, you need not worry; it took the best mathematicians in the human race more than two centuries to realize that such a thing was possible. Further, when I do give you an example, you may feel that this example is rather unfair. That also is to be expected; evidently a curve that passes through a point, but does not pass through it in any direction, is something rather peculiar; we can hardly expect it to be just like the graphs we did in elementary algebra.

How should we go about making up an example of a curve that does not have a direction at a particular point? If we had such a curve and we tried to find the slope of the curve at this point, we should not arrive at any answer. I do not mean just that we should not be able to calculate the slope; I mean that there would be no slope to calculate. How could such a situation arise? To answer this, we need to recall how we went about finding y', which measures the slope of a curve. This was done on pages 53–56. There, we had the diagram of Fig. 61.

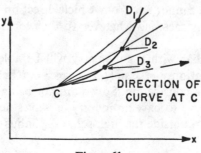

Figure 61

We took points $D_1, D_2, D_3, \ldots$ on the curve, and worked out the slopes of the lines $CD_1, CD_2, CD_3, \ldots.$ In the examples we considered, we found that these slopes approached a fixed number, which we called the slope of the curve at C. We did not prove that these slopes must approach a fixed number; we just observed, in particular cases, that they did so. Suppose, however, that they do not settle down, but wander about endlessly. Can this happen? If so, what would the curve look like?

Imagine a curve like the one shown in Fig. 62. The dashed lines make

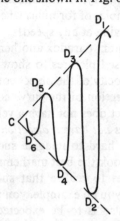

Figure 62

angles of 45° with the horizontal. The points D_1, D_2, D_3, D_4, D_5, $D_6, \ldots$ are chosen approaching closer and closer to C. But the slopes of the lines CD_1, CD_3, CD_5 are $+1$ while the slopes of the lines CD_2, CD_4, CD_6 are -1. Thus the slope waves back and forth between $+1$ and -1 as D approaches C. *We suppose this to continue indefinitely.* This means, of course, that the curve must be very complicated near C. There must be an infinity of hilltops and an infinity of valley bottoms in the neighborhood of C. Then, as D approaches C, the line CD continually oscillates between the two dotted lines; its slope never settles down towards any particular value. The curve is approaching the point C, but we cannot say from which direction it approaches. We cannot attach any meaning whatever to the words "the slope of the curve at C."

Someone familiar with trigonometry will be able to verify that the graph of the equation $y = x \sin(1/x)$ near $x = 0$ behaves like the curve we have just been considering. So we are not being particularly unfair in requiring a curve to have an infinity of waves near the point C; we do not need to go outside the high-school syllabus to give an example of such a curve.

We can also construct such a curve without using trigonometry. The construction should be clear from Fig. 63 and the description that follows. *CA* can be of any convenient length. *B* is halfway from *C* to *A*. The triangle *ABE* is right-angled with angles of 45° at *B* and *A*. A quadrant of a circle can thus be drawn from *A* to *B*, with center *E*. The point *G* is halfway from *C* to *B*. The triangle *GFB* is like the triangle *AEB*, but to half-scale and the other way up. *B* and *G* are connected by a quadrant of a circle, center *F*. Triangle *IHG* is obtained by drawing *GFB* to half-scale, and a quadrant of a circle is drawn with center *H*.

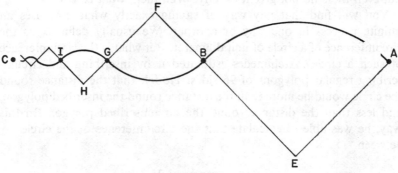

Figure 63

This construction is continued indefinitely, so that we have circle quadrants alternately above and below the line *CA*, each quadrant being one half the scale of the previous one.

If we consider the sequence of points, *A, B, G, I, ..* , each of these is half the distance from *C* that the previous one was. Thus we can continue our construction indefinitely, getting always nearer to *C*, but never passing it.

If a point *D* moves along this curve towards *C*, the line *CD* oscillates very much as it did in Fig. 62. The slope of *CD* takes positive and negative values alternately as *D* approaches *C*. It can be shown that these values range repeatedly from +1/7 to −1/7 and back again. So the slope never settles down to any particular value as *D* approaches *C*.

I expect you will raise certain objections to this example. (1) You may object that this curve is not really constructed at all, because an infinity of circles have to be drawn, and it would require an eternity to reach *C*, the point we are interested in. (2) In any case, you may say, this is not *a curve*; it is lots of bits of different curves stuck together.

In making these objections, you are in good company. The second objection was made by some very great mathematicians a century or two ago. The first objection is of a kind that still arouses strong feelings and fierce arguments between mathematicians.

Your second objection will be considered later on. We take a look at your first objection. It is natural that you should see logical difficulties in a construction that takes an eternity to carry out. But have you thought what is involved if we are only allowed to speak of things that can be constructed in a finite number of steps? I imagine you speak, from time to time, of the number π. Will you tell me exactly what number π is? Someone may suggest that it is $3\frac{1}{7}$. Is it exactly $3\frac{1}{7}$? No; it is more like 3.1416, which is rather less than $3\frac{1}{7}$. Is it exactly 3.1416 then? No; it has been calculated to thousands of places of decimals, but even these do not give it exactly. Well then, what is it?†

You will find that any way of saying exactly what π is uses an infinite process in one way or another. We usually define π as the circumference of a circle of unit diameter. But what is the circumference of such a circle? Archimedes estimated it by inscribing and circumscribing regular polygons of 96 sides. He felt that the distance round the circle would be more than the distance round the inscribed polygon, and less than the distance round the circumscribed polygon. In this way, he was able to calculate that the circumference of the circle lay between

$$3\frac{10}{71} \quad \text{and} \quad 3\frac{1}{7}.$$

But why stop at polygons of 96 sides? By taking more sides, you could get more accurate estimates. But the process is unending. At no actual stage is it complete; at any moment, we can only say that π lies between certain numbers. The exact number π is only defined by taking *all* these estimates; each estimate gives an interval within which π must lie; π is the only number that lies in all these intervals; it is the only number that is larger than the perimeter of any inscribed regular polygon, and less than the perimeter of any circumscribed regular polygon.

There is another way of calculating π. This is a purely arithmetical procedure. In many ways, it is simpler than the geometrical argument used above. It has the disadvantage that I cannot explain the proof to you. However, within your first year of studying calculus you could come to understand how this method was devised. Here then is the method. Consider the series

$$1 - \frac{1}{3} + \frac{1}{5} - \frac{1}{7} + \frac{1}{9} - \cdots.$$

† You will find a table of π to 4,000 places of decimals in *The Lore of Large Numbers* by P. J. Davis, published in this series.

Let S_n stand for the sum of the first n terms of this series. Thus

$$S_1 = 1, \qquad S_2 = 1 - \frac{1}{3}, \qquad S_3 = 1 - \frac{1}{3} + \frac{1}{5},$$

and so on. It can be shown that if you take an odd value for n, then S_n will be larger than $\frac{1}{4}\pi$. If you take an even value for n, then S_n will be less than $\frac{1}{4}\pi$. Further, this statement fixes the value of $\frac{1}{4}\pi$. There is no other number that is smaller than each one of the numbers S_1, S_3, S_5, ..., and larger than each one of the numbers S_2, S_4, S_6, ... But it will take us an eternity to calculate all of these numbers. So this method also does not allow us to define $\frac{1}{4}\pi$ exactly, without supposing an infinite process to have been completed. It does, of course, allow us to determine $\frac{1}{4}\pi$ *as closely as we like*; by calculating the sum of the first million terms of the series, we could determine $\frac{1}{4}\pi$ to about five places of decimals; multiplying this result by 4, we should obtain an estimate of π. But no such finite procedure ever gives us π exactly.

Actually, the series is not a very good one for the practical purpose of calculating estimates of π. However this is not very important for our present purpose, which is to show that no one has yet found a method of specifying π exactly without appealing to some infinite process.

Accordingly, if you are going to object to my construction for the curve on page 97 because it involves an infinity of steps, you have got to object to a lot of other things as well! You must object whenever anyone mentions the number π, or the length of the circumference of a circle, or the area of a circle. All of these can be defined only in terms of an unending process. Calculus, in fact, is essentially concerned with unending processes. The true speed, s', was something continually approached, but never reached, by the average speed over a small interval of length h. The slope of a curve, y', was something continually approached but never reached by the slope of the line CD. If we are to allow unending processes in the finding of slopes and speeds, why should we exclude them in the construction of curves and laws?

Even in arithmetic, an unfinished process occurs. If you try to express the fraction $\frac{1}{9}$ as a decimal, you obtain the result $0.1111111 \cdots$ which contains an unending sequence of ones. In arithmetic, we usually write quite cheerfully $\frac{1}{9} = 0.111111 \cdots$, but really some explanation is called for when we say that an unending expression is equal to $\frac{1}{9}$. What do we mean by such a statement? How can we test it? We can explain our meaning in the following way. Let

$$S_1 = 0.1,$$
$$S_2 = 0.11,$$
$$S_3 = 0.111,$$

and so on. S_n will stand for the decimal that has the figure 1 occurring n times after the decimal point. In the sequence S_1, S_2, S_3, ..., each number is nearer to $\frac{1}{9}$ than the previous one, and by going far enough along, you can make the difference as small as you like. For, in fact,

$$9S_1 = 0.9 \quad = 1 - 0.1,$$
$$9S_2 = 0.99 \quad = 1 - 0.01,$$
$$9S_3 = 0.999 = 1 - 0.001.$$

By choosing a sufficiently large value of n, we can make $9S_n$ as close to 1 as we wish. So, as n gets large, $9S_n$ gets closer and closer to 1, and this means that S_n is getting closer and closer to $\frac{1}{9}$. Accordingly, the value $\frac{1}{9}$ is picked out from all other possible values when we write down the numbers S_1, S_2, S_3, ... This sequence settles down to the value $\frac{1}{9}$ and to no other value.

You will notice that I have used the same idea to explain what I mean by the unending expression 0.11111... as I did a little earlier to explain what I meant by the unending expression

$$1 - \tfrac{1}{3} + \tfrac{1}{5} - \tfrac{1}{7} + \cdots.$$

Any time in future that I use an unending expression, we will agree that this is how it is to be interpreted. We break off the expression at a certain point and calculate the value; we then see whether this value approaches any fixed number when more and more terms are included in the calculation.

The unending decimal 0.11111 $\cdots$ could be written as

$$\left(\frac{1}{10}\right) + \left(\frac{1}{10}\right)^2 + \left(\frac{1}{10}\right)^3 + \left(\frac{1}{10}\right)^4 + \cdots.$$

The fraction $\frac{1}{10}$ is no better than other fractions. We might find interesting results by considering the unending expression

$$\left(\frac{2}{3}\right) + \left(\frac{2}{3}\right)^2 + \left(\frac{2}{3}\right)^3 + \left(\frac{2}{3}\right)^4 + \cdots.$$

We could form and investigate many other expressions of this type. To avoid the labor of investigating each one separately, we may use algebra and do the whole lot at once by investigating the unending expression

$$x + x^2 + x^3 + x^4 + \cdots.$$

This is in fact a standard topic of high-school algebra, where it is shown that, provided x is a proper fraction, the above expression settles down to the value $x/(1 - x)$. You may notice that, if you substitute $\frac{1}{10}$ for x, this formula does give you the result $\frac{1}{9}$ in agreement with our earlier work.

The unending expression $x + x^2 + x^3 + \cdots$, where x is any proper fraction, can thus be replaced by the ordinary algebraic expression $x/(1 - x)$, so that we have not obtained anything new. However, in many cases an unending expression gives us *something that could not be got in any other way*. For example, the expression,

$$m - \frac{m^3}{3} + \frac{m^5}{5} - \frac{m^7}{7} + \frac{m^9}{9} - \cdots,$$

where m is any proper fraction, cannot be replaced by any ordinary algebraic expression. It gives us the length s in Fig. 64. In this figure, APB is part of a circle of radius unity with center at the origin O. The line $y = mx$, of slope m, cuts the circle in the point P. The arc AP has the length s.

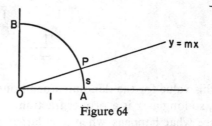

Figure 64

There are, in fact, many very interesting laws that can only be expressed with the help of infinite series. These have been very successfully investigated with the help of calculus. Both in pure mathematics and in science, the results so obtained have proved extremely reliable and satisfactory. Mathematics would be poorer and science would be paralyzed if infinite processes were outlawed. For these reasons, mathematicians continue to use unending constructions in their work. But we do so with the knowledge that we are handling dynamite; infinity can be used, but it must be used with care.

This discussion arose from an attempt to allay any doubts that you might feel about the unending construction used on page 97. We called this objection (1). But the discussion also throws light on objection (2), that the graph in Fig. 63 is not a single curve, but is obtained by piecing together bits of different circles.

Consider the following equation,

$$y = \frac{1}{1 + x + x^2 + x^3 + x^4 + \cdots}$$

where x is supposed to be positive. I think you will agree that this is *one* equation, and that its graph is accordingly given by one formula. What is the graph of this equation?

First of all, consider the part of the graph corresponding to values of x between 0 and 1. Since x is a proper fraction, by the result given on page 101,

$$x + x^2 + x^3 + \cdots = \frac{x}{1 - x}.$$

Adding 1 to both sides, we find

$$1 + x + x^2 + x^3 + \cdots = 1 + \frac{x}{1 - x}$$

$$= \frac{1 - x}{1 - x} + \frac{x}{1 - x}$$

$$= \frac{1}{1 - x}.$$

Substituting this value for the denominator in our equation, we find that $y = 1 - x$, so long as x is a proper fraction.

Now let us see what happens when x is larger than 1. If we take $x = 2$, for example, we find

$$y = \frac{1}{1 + 2 + 4 + 8 + 16 + \cdots}.$$

We find the value of y by taking n terms of the series in the denominator, and seeing what number the fraction approaches as n gets larger and larger. For example, if we take five terms in the denominator, the fraction equals 1/31; if we take ten terms, the fraction becomes 1/1023; if we take twenty terms, it becomes 1/1,048,575. The more terms we take, the smaller the fraction becomes; in fact, it is approaching zero, and this is the value of y for $x = 2$. If you take any other number larger than 1, you will be led to the same conclusion. For all x larger than 1, the value of y is zero.

Thus this single formula manages to give us pieces of two algebraic graphs. For x between 0 and 1, the graph coincides with the line $y = 1 - x$. For x larger than 1, the graph coincides with $y = 0$. Thus the equation has the graph of Fig. 65.

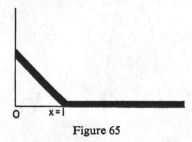

Figure 65

With the simple equations studied in elementary algebra, it is impossible to get this kind of effect. But when infinite series are allowed, graphs which seem to consist of a number of separate geometrical figures become quite commonplace.

In electronics and other branches of science, a special type of unending expression is used. It is called a Fourier series. With a Fourier series you can quite easily obtain the graphs of Fig. 66.

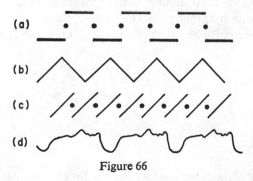

Figure 66

Graph (c) of the figure is of importance for time bases in television and radar. It represents the type of motion in which a spot of light moves across a screen at a steady pace, then suddenly goes back to the beginning and starts all over again.

Curve (d) is a little joke which I found in a book on the theory of music. Someone took a photograph of a friend, and fed the profile into a machine known as a harmonic analyzer. The machine computed a formula, the graph of which is the outline of that face repeated again and again.

Graph (a) is sometimes called the graph of the *Parapet Function* and graph (b) the *Sawtooth Function*.

We do not need to go to electronics for examples of graphs that seem to consist of many different curves. Figure 67 shows the behavior of a bouncing ball.

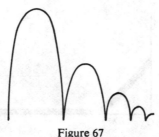

Figure 67

It is interesting that—in theory at any rate—a ball makes an infinite number of bounces in a finite time. For example, at each bounce it may rise to one-fourth the height of the previous bounce, and each bounce takes half as long as the previous one.

A Curve with No Direction Anywhere

In Fig. 62 we drew a curve which passed through the point C, but had no direction at that point. Of course, the same could be said of a much simpler curve like that of Fig. 68. This curve has a sudden bend

Figure 68

at the point P, and we cannot draw a tangent at that point. There is nothing very surprising in this figure. The curve simply enters P from one direction and leaves it in another.

In both these examples, there is a single point where the curve mis-behaves. In Fig. 62, the curve wobbles at C; in Fig. 68, the curve has a bend at P. But at other points the behavior is entirely normal. For several centuries it was thought that wobbles and bends must be exceptional things, that could only occur at isolated points. However in 1875 a paper was published showing that you could have a curve consisting of nothing but wobbles. You can choose any point on this curve you like; the curve passes through that point but does not pass through it in any direction! This surprised mathematicians very much. It became clear that in calculus we should not begin by asking, "What is the slope of this curve at the point P?" but rather, "Does this curve have a slope at the point P?"

You may feel that such a curve is too painful to think about. Let us shut it out, and say we will not think about it. In practice, we do this to some extent in first-year calculus. With a simple formula like $y = 2x^3 - 3x^2$, we ask the student to find y'; we do not, as a rule, emphasize the question of whether y' exists at all. At this stage in calculus, we are dealing with simple formulas, for which y' is bound to exist; there may of course be points like the origin on the curve $y = \sqrt{x}$, where the tangent is vertical so that no finite value exists for y'. Still, the curve does have a definite direction. In our experience of high school graphs, overwhelmingly we meet smooth, well-behaved curves.

So we might try to deal with the situation by outlawing the curves that have no direction. We might say that it is unfair to define such a curve, and we will not accept such a definition. There are several reasons why we do not do so. First of all, it would be cowardly. We have come to a region where things behave differently from what we have been used to; shall we turn back and go home? The mathematician has the instincts of the explorer; at all costs, go forward; if things are different, so much the better; that will make it more interesting.

There are other, more definite reasons. This new, strange region borders our own countryside. Often we shall be chasing some mathematical quarry and it will cross the boundary line. We do not want to give up the chase at that point. For in fact the strange curves are defined by formulas which look exactly like those we use, not only in pure mathematics, but also in engineering and science. They can be defined by Fourier series, which are of the utmost importance in science and indeed owe their origin to mathematical physics. In life, it is hardly ever possible to draw a boundary line between what you will study and what you will not; everthing combines to push you across such an artificial barrier.

Suppose, then, we decide to think about directionless curves. We do not merely say that such curves exist. We can produce an actual formula and know that this formula gives a directionless curve.† What would happen if we tried to draw the graph of a directionless curve? Something like the following: Suppose we first decide to calculate the values of y corresponding to the whole numbers $x = 0, 1, 2, 3$. We

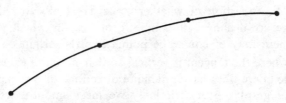

Figure 69

might get points like those of Fig. 69, and we would think, "The curve must run like this."

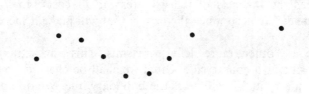

Figure 70

To make sure, we decide to plot some more points. So we calculate y for values of x at intervals of $\frac{1}{4}$, say for $x = \frac{1}{4}, \frac{1}{2}, \frac{3}{4}, 1, 1\frac{1}{4}, \ldots$. We find they lie as in Fig. 71. So we revise our ideas of what the curve

Figure 71

† J.L.B. Cooper, in the article "Mathematical Monsters," *Mathematical Gazette* (December, 1954) gives the example

$$f(x) = \sin x + \frac{1}{4} \sin 2x + \frac{1}{9} \sin 6x + \frac{1}{16} \sin 24x + \cdots.$$

The general term is

$$\frac{1}{n^2} \sin n! \, x.$$

looks like. It now seems that it should be something like this:

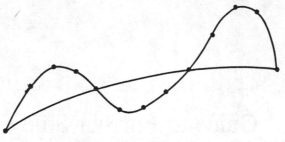

Figure 72

Again we plot some more points, and we find these indicate yet more waves. The extra points (at intervals of $\frac{1}{16}$) give us Fig. 73 and suggest

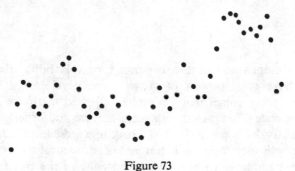

Figure 73

the curve of Fig. 74. And so it goes on. At each stage, we find evidence

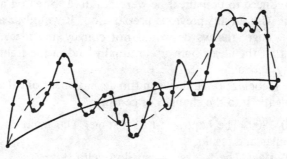

Figure 74

of shorter and shorter waves. The curve is infinitely crinkly! And yet it is a perfectly good, definite curve. We can work out as many points on it as we like, just as when we plot an elementary graph. The more points we plot, the more clearly do we see how the curve lies. But we cannot sketch the curve with a sweep of a pencil as we do with simpler graphs.

Guide to Further Study

As was emphasized at the beginning of this book, the ideas of calculus have great powers of growth and development; from this small root there come many branches of pure mathematics and of physical science. This growth will seem natural and orderly to anyone who traces its development from the root upwards. But the latest fruits of this growth may seem very strange and unnatural to someone who meets them suddenly without any knowledge of the tree from which they came. It is therefore extremly important to read books about calculus in *the correct order*. A student with real genius for mathematics might be reduced to despair, if he were required to read a modern text on analysis without any previous preparation. It would be a book in a foreign language; the words would not convey any ideas. This does not mean that the ideas, built up gradually and in the right order, are particularly difficult.

One can recognize three stages in the development of calculus, which fit rather neatly into the changes of century.

(1) 1600–1800. The happy-go-lucky stage. The main emphasis is on formulas and results.

(2) 1800–1900. The analysis or epsilon-delta stage.

(3) 1900–   . The stage of abstraction and extreme generalization.

In passing from each stage to the next, new ideas and a new way of thinking have to be learned. A student may experience some kind of crisis. At first, he feels he cannot grasp these new ideas. If he keeps reading about them in different books, and thinks and works problems for himself, he should reach a stage where everything seems to fall into place; he then finds it hard to see why these ideas ever seemed difficult to him. He sees that the new ideas are simply the old ideas expressed in a different way, perhaps a little more clearly.

Before 1900, there was a general belief that calculus was much too difficult to teach young mathematicians. Many results that could have been proved very simply by calculus had to be obtained by more painful algebra. This procedure was known as "calculus dodging." Round about 1900, John Perry and others in England began to advocate the view that the essential ideas and methods of calculus were simple and could be taught in schools. E. H. Moore, professor of mathematics at Chicago and the father of modern American mathematics, gave this view his blessing.† One of Moore's students, F. L. Griffin, pioneered the teaching of calculus to first-year college students. This was regarded as a very daring thing to do. Griffin's celebrated book, *Introduction to Mathematical Analysis*, shows how he went about this. The word "calculus" did not appear in the title, in case the students were terrified. This book, which deals both with trigonometry and calculus, and emphasizes the relation of calculus to physics and engineering, can be warmly recommended as an introduction to calculus.

In England, during the past fifty years, the teaching of calculus in high schools has become the general practice. It is not possible to say just how many years of calculus are done in school as, in the better schools, students are encouraged to work ahead at their own pace. Whether a student meets calculus at 18 or 16 or 14 depends largely on his own ability. As there are no books written in America for teaching calculus to 15-year-olds, it may be of interest to mention some English calculus texts. Being much smaller and more concise and plainer than American texts, these books are very inexpensive.

Introductory calculus ideas are usually brought in towards the end of the algebra text. See, for example, Durell, Palmer, and Wright, *Elementary Algebra* (Bell, Portugal Street, London, W.C.2).

Fawdry and Durell, *Calculus for Schools* (Arnold, London) gives a very simple introduction to calculus.

Durell and Robson, *Elementary Calculus*, volumes I and II (Bell), introduces calculus in a simple way; the authors make great efforts not to make any statement that the student will find to be untrue when he reaches a more advanced stage. This book takes the student further into calculus than *Calculus for Schools* does. Volume II explains, among other things, the idea of *partial differentiation*, which is of importance for further mathematics and, in particular, for mathematical physics.

† See the first yearbook of the National Council of Teachers of Mathematics (U.S.A.).

However, you will need to work far more exercises than are given in this book, if you are to remember what you read. You should seek out exercises from any book on calculus† and work at them in such a way that you always keep in good practice.

Piaggio, *Differential Equations* (Bell) is a very readable book that might follow *Elementary Calculus*, Part II. The earlier chapters, in particular, do not delve into underlying theory but give the student an idea of how calculus is used. Chapter IV gives a very quick and simple introduction to Fourier series.

We now wish to pass from the happy-go-lucky stage to the epsilon-delta stage. Most texts do this far too suddenly. The book that takes the student most gradually and carefully from the old to the new viewpoint seems to me to be Hardy, *Pure Mathematics* (Cambridge University Press).

You may wonder why we call this the "epsilon-delta" stage. In the 19th century, many ideas which had been previously accepted as sufficiently clear—for example, "continuous", "approaches"—were carefully analyzed and defined. The new definitions usually contained the phrase "given any positive ϵ, however small, δ can be found such that ...".‡ People came to think of this phrase as typical of the new analysis.

It may help you to acquire these new ideas if you read some general accounts of how mathematics developed in this direction; for example:

Tobias Dantzig, *Number, the Language of Science* (Doubleday Anchor, 95 cents). Especially, Chapters 7, 8, 9.

Felix Klein, *Elementary Mathematics from an Advanced Viewpoint; Arithmetic, Algebra, Analysis* (Dover).

W. W. Sawyer, *Mathematician's Delight* (Penguin, 85 cents).

Once you have reached the stage where you can read a book written in the epsilon-delta language, there is no doubt what your next book should be—Courant, *Differential and Integral Calculus* (Interscience, N.Y.). This book is admirably clear. As Nathan G. Park says, in his *Guide to the Literature of Mathematics and Physics*, "Courant will give the student the best possible balance between vigor and rigor."

Where you go after this must depend very much on your personal tastes and aims. There is so much mathematics that, unfortunately, no one can learn the whole of it.

---

† A bad book can contain good exercises. For example, J. Edwards, *The Differential Calculus* (St. Martin's Press), is famous for the number of illogical and untrue statements it contains. But it contains an amazing collection of examples for anyone wishing to master formal manipulation in calculus.

‡ The symbols δ, ϵ are read "delta" and "epsilon". They are the letters of the Greek alphabet corresponding to our *d* and *e*.

In every branch of mathematics there are a few central ideas; in the following-out of these central ideas, all kinds of detailed investigations become necessary. Very many books give you the details without the central ideas that illuminate the whole subject. You should not therefore be disturbed if, when trying to learn a new branch of mathematics, you find the books on it completely incomprehensible. Continue to search in libraries or bookshops until you find a book that gives you the essential ideas. Sometimes you cannot find one book that gives you all you want; you may have to pick up a clue here and a clue there.

Some of the mathematics of the 20th century, in particular, seems most strange if you are suddenly plunged into the middle of it. It appears to be an entirely different subject from the mathematics you learned at school. Yet it grew from the older mathematics. This happened in something like the following way. The older mathematics dealt in the main with definite objects. You had to solve a particular equation, or prove a theorem about some particular shape in geometry, or study the vibrations of a particular mechanical system. As time passed, more and more special results about particular objects accumulated, and mathematicians began to long for some way of systematizing the subject. There were too many details for anyone to remember them all. Then it began to be noticed that, very often, the details were only obscuring the picture. Of all the information available about some object, only a small part might be necessary for solving the problem in hand; that aspect was helpful, all the rest was merely distracting. Mathematicians began to study these special aspects, much as a chemist might extract a vitamin from a complex substance. Someone who knew nothing about vitamin pills might not realize that such a thing was food at all. In the same way, a person new to modern abstract mathematics might not realize that it was mathematics at all.

This extraction of the essential ideas was also made necessary by mathematicians going on to more and more complicated problems. Some vibrations in mechanics can be represented by the motion of a point in two or three dimensions. We are able to visualize the mechanical problem by means of geometry. Some more complicated problems require four or five or six or more dimensions to visualize. So we develop the geometry of n dimensions, and this helps us to visualize the problem, in a somewhat vaguer way, by the analogy with ordinary space of 3 dimensions. Some problems require an infinity of dimensions. Now space of infinite dimensions in some ways resembles space of three dimensions, and in some ways differs from it. So it becomes necessary to separate very carefully those ideas we have about ordinary geometry which are still true and helpful when we are thinking about infinite dimensional space, from those which are untrue and misleading when used as analogies. By this kind of road mathematicians reached the concept of *Hilbert space*.

The connection between physics and the geometry of space is brought out very nicely in Courant and Hilbert, *Methods of Mathematical Physics* (Interscience, N.Y.), Volume I.

A book which carries the reader from 19th- to 20th-century mathematics, without any sense of a sudden break, is Riesz and Nagy, *Functional Analysis* (Ungar, N.Y., 1955).

By contrast, one may mention Munroe, *Introduction to Measure and Integration*. This book from the start has the flavor of the 20th century. For a reader with the necessary background it is extremely clear.

E. J. McShane, *Integration* (Princeton) is written for "students of little maturity" who are beginning graduate work in mathematics. Any student of strong mathematical ability will, of course, be able to read it several years earlier than this.

A student who finds difficulty in passing from traditional calculus to the set-theoretical approach may find something of interest in the huge book, Hobson, *Functions of a Real Variable* (Cambridge University Press, reprint by Dover). This book has been described as a strange mixture of careful rigor and astonishing errors. It was written in the years when the new theories were coming in, so you see Hobson (who had grown up under the older approach) trying to explain to himself and others what these new ideas are. It is a book to browse in, rather than to read from cover to cover. The fact that the book contains errors† is valuable. It means that you cannot accept any statement on authority; all the time, you have to ask yourself, "Do I believe this?"

† See Littlewood, *A Mathematician's Miscellany* (Methuen, London), page 68.

List of Technical Terms

Throughout this book, I have explained things as far as possible in everyday language. When you read other books on calculus, you will need to know the symbols and the special names that mathematicians use.

Derivative. s' is called the derivative of s. You may also find the symbols ds/dt, Ds, $D_t s$ used for the derivative. These have exactly the same meaning as s'.

Differentiation. The problem of finding the derivative is called *differentiation*. Thus in Chapter 3 you learned how to differentiate t^2, in Chapter 4 how to differentiate t^n, and in Chapter 5 how to differentiate any polynomial.

Integration. Finding an area or a volume is a problem of *integration*. Integration can be regarded as the reverse of differentiation. The symbol $\int$ is used in connection with integration. At the end of Chapter 9 we found the volume of a half sphere. A mathematician would write our result

$$\frac{2}{3}\pi = \int_0^1 \pi(1 - t^2)dt.$$

Limit. Many times in this book we have noticed that something "approached" or "seemed to be settling down to" a certain value. In Chapter 2, the numbers 5, 5.9, 5.99, 5.999, ... seemed to be approaching the value 6. In Fig. 23 on page 54, the slope of the line

CD approached closer and closer to the slope of the curve at *C*. By writing enough ones, you can make $0.11111 \cdots 111$ approach as close as you like to the fraction 1/9. In each of these cases, something is tending *towards a limit*. While the word has not been stressed, the idea of limit runs through everything discussed in this book.

Function. The meaning of the word *function* has developed and changed during the last three centuries. At first, "*y* is a function of *x*" meant something very much like "*y* is related to *x* by some formula." This would cover, for example, $y = 2x + 1$ or $y = x^2$ or $y = \sqrt{x^3 + 1}$. In each of these cases, an 18th-century mathematician sees a formula giving *y* in terms of *x*. Suppose he has some procedure that can be applied to each of these, and to many other formulas as well. He does not mind what the particular formula is; he wants to lump them all together. He would say, "Let *y* be any function of *x*." He would write this, for short, as $y = f(x)$.

As time passed, this viewpoint proved insufficient. In Fig. 65, we had a graph consisting of parts of two lines. Between $x = 0$ and $x = 1$, the value of *y* was $1 - x$. For *x* larger than 1, the value of *y* was zero. So two formulas were involved, $y = 1 - x$ and $y = 0$. What shall we say? Do we have two functions here, or part of one function grafted onto part of another function, or what? There were furious discussions between mathematicians about this question. As time passed, more and more strange graphs came to the attention of mathematicians, and it was eventually decided that the best thing to do was to forget all about the simple formulas of algebra. Instead, it was decided to write $y = f(x)$ if any procedure whatever fixed the value of *y* so soon as the value of *x* was given. Thus the graph of Fig. 65 defines a function; if I tell you any positive number for *x*, you can read the corresponding value of *y* from the graph. If I say $x = 2$, you answer $y = 0$. If I say $x = \frac{3}{4}$, you answer $y = \frac{1}{4}$. You are never at a loss for an answer. As soon as I say the value of *x*, that fixes the value of *y*. Good; we do not inquire any further into the matter. Any procedure that associates a single value of *y* with each value of *x* defines a function.

The graph of Fig. 65 was drawn only for positive values of *x*. So the function is not defined for all values of *x*, but only for positive values. Mathematicians have decided that this is nothing to worry about. In first-year algebra, $y = \sqrt{x}$ is defined only for positive *x*. We do not know anything about the square root of a negative number. We accept this situation. We say that $\sqrt{x}$ is defined (in beginning algebra) only for the *domain* of positive values of *x*. If $y = f(x)$ is defined only for a certain set of values of *x*, these values are said to form the *domain of the function*.

For example, if x is a whole number, we can define y as the largest prime factor of x, but this definition would not make sense if x was a fraction. We have defined a function for the *domain* of the whole numbers.

In traditional algebra, x and y stand for numbers. But functions can be defined which have nothing to do with numbers. For example, suppose we consider all the instants of time since the year 1789. These form the domain of the function. At any instant during those years, the President of the United States had eyes of some color. With some historical research, one could find out what that color was, corresponding to each time. So we have a procedure for associating a definite color with each instant of time since 1789. This procedure defines a function. We have come a long way from algebraic formulas! The word "function" is generally used today in this very wide sense.

One point arises. Returning to ordinary algebra, we might consider the following two procedures.

Procedure I: Take any number, x.
     Add 1 to it.
     Square the result.
     This gives the value of y.

Procedure II: Take any number, x.
     Square it.
     Add twice the number.
     Add 1.
     This gives the value of y.

Each procedure defines a function. The procedures are different. Shall we say that the resulting functions are different?

If we take for example $x = 5$, Procedure I gives $y = (5 + 1)^2 = 36$ and Procedure II gives $y = 5^2 + 2 \cdot 5 + 1 = 36$. So either procedure leads us to associate $y = 36$ with $x = 5$. And the same happens, of course, with any number you may choose. Procedure I corresponds to the formula $y = (x + 1)^2$, and Procedure II to the equivalent formula $y = x^2 + 2x + 1$.

Mathematicians have agreed to say that both procedures define the *same* function. We are only interested in the final result, not in the details of the calculation. If we have any procedure, which leads you to say $y = 36$ when I say $x = 5$, and makes you say $y = 4$ when I say $x = 1$, and quite generally makes you say $y = (n + 1)^2$ when I say $x = n$, then that procedure defines the same function as Procedure I above.

You may meet a definition of function which begins: "A function is a set of ordered pairs...". This is a very condensed and abstract way of saying what I have outlined above. I am not too happy myself with any definition that begins "A function is...", any more than I should be happy with a definition that began "Electricity is...", or "Magnetism is...", or "Gold is...". I can give you a series of tests, in each case, that will enable you to say, "This is probably an electrically charged object", or "This is probably a magnet", or "This is probably a piece of gold". In the same way, I have given tests above that will enable you to tell (i) whether a particular procedure defines a function, and (ii) whether two apparently different procedures define the same function.

It is important to distinguish between the *function* and the *value of the function*. If f stands for the function defined by Procedure I above, we may write $36 = f(5)$. This means that 36 is the value of y that Procedure I leads us to associate with $x = 5$.   36 is called the *value of the function* for $x = 5$. It would be wrong to say that 36 *is the function f*. It would be nearer the truth to say that the letter f, by itself, indicates the operation of "adding 1 and then squaring". $f(5)$ represents the result when this operation is applied to the particular number 5.

Answers to Questions and Exercises

p. 12 1. The steeper the line is, the faster the object is moving.

p. 13 2.   (a) is (ii)   (d) is (iv)
                           (b) is (iii)   (e) is (i)
                           (c) is (v)

      3.   (f) is (viii)   (h) is (ix)
                           (g) is (vi)   (i) is (vii).

p. 20 1. $s = 20t.\ s' = 20.$

      2.

t	0	1	2	3
s	0	30	60	90

                  Velocity is 30 mph. $s' = 30.$

      3. $s' = 40.$

      4. 50.

      5. $k.$

p. 20 (1) 10   (2) 10   (3) 10
        If $s = 10t + c,\ s' = 10.$

p. 21 (4) 20   (5) 20   (6) 20   (7) 20
        Conclusion: If $s = 20t + c,\ s' = 20.$

        (8) 30   (9) 50   (10) 40   (11) 30   (12) 50
        The illustrations are straight lines. If the scale is kept fixed, the larger the velocity s' is, the steeper the line will be. If different laws give the same velocity, as in (1), (2), (3), the lines will be parallel.

117

p. 33 1.

t	1	1.001	2	2.001	3	3.001	4	4.001	5	5.001
s	1	1.003	8	8.012	27	27.027	64	64.048	125	125.075

t	1	2	3	4	5
s'	3	12	27	48	75

Law: $v = s' = 3t^2$.

2. See p. 34.

3. See p. 34.

4. $5t^4$; $6t^5$; nt^{n-1}. See pp. 34–35.

p. 44 (1) $20t$ (4) $800t^{99}$

(2) $60t^2$ (5) $6t^2$

(3) $16t^3$ (6) $6t$

p. 45 (1) $20t + 60t^2$ (4) $35t^6 - 8t^3$

(2) $6t^2 + 6t$ (5) $20t + 60t^2 - 20t^3$

(3) $35t^6 + 8t^3$

p. 47 1. $0; 2; 2$ 5. $10t - 4$

2. $0; 3t^2; 3t^2$ 6. $6t^2 - 6t - 10$

3. $0; 3; 2t; 2t + 3$ 7. $80t^{19} + 30t^{14} - 30t^9 + 5$

4. $10t + 4$ 8. $60t^5 + 60t^4 - 60t^3 + 60t^3 - 60t + 60$

p. 52 (1) 1   (2) 1   (3) 2   (4) 3

p. 53 (5) -1 (6) -2

p. 68 6. $3x^2 - 6x + 9 = 3(x - 1)^2 + 6$, never zero and never negative. If $y = x^3 - 3x^2 + 9x$, $y' = 3x^2 - 6x + 9$. So y' is always
positive. Curve uphill; resembles Fig. 37.